Choosing Safety

A Guide to Using Probabilistic Risk Assessment and Decision Analysis in Complex, High-Consequence Systems

Michael V. Frank

New York • London

711 Third Avenue, New York, NY 10017, USA
2 Park Square, Milton Park, Abingdon, Oxon, OX14 4RN

Library of Congress Cataloging-in-Publication Data

Frank, Michael V.
Choosing safety : a guide to using probabilistic risk assessment and decision analysis in complex, high-consequence systems / Michael V. Frank.
p. cm.
Includes index.
ISBN 978-1-933115-53-5 (hardcover : alk. paper) -- ISBN 978-1-933115-54-2 (pbk. : alk. paper)
1. Technology--Risk assessment. 2. Risk management. I. Title.
T174.5.F725 2007
658.4'04--dc22

2007028657

The findings, interpretations, and conclusions offered in this publication are those of the author. They do not necessarily represent the views of Resources for the Future, its directors, or its officers.

ISBN 978-1-933115-53-5 (cloth)
ISBN 978-1-933115-54-2 (paper)

About Resources for the Future and RFF Press

Resources for the Future (RFF) improves environmental and natural resource policy-making worldwide through independent social science research of the highest caliber. Founded in 1952, RFF pioneered the application of economics as a tool for developing more effective policy about the use and conservation of natural resources. Its scholars continue to employ social science methods to analyze critical issues concerning pollution control, energy policy, land and water use, hazardous waste, climate change, biodiversity, and the environmental challenges of developing countries.

RFF Press supports the mission of RFF by publishing book-length works that present a broad range of approaches to the study of natural resources and the environment. Its authors and editors include RFF staff, researchers from the larger academic and policy communities, and journalists. Audiences for publications by RFF Press include all of the participants in the policymaking process—scholars, the media, advocacy groups, NGOs, professionals in business and government, and the public.

Contents

Contents

Preface

OVER THE LAST CENTURY we have seen the catastrophic consequences of numerous failures resulting from human actions—failures ranging from buildings and bridges to space and launch vehicles, from chemical factories to nuclear power plants, from ships to airplanes, and from trains to automobiles. Often the root cause of these failures can be traced to decisions made during their design and engineering. The ideas, methods, and case studies I present in this book are at the nexus of probabilistic risk assessment (PRA) and decision analysis. In these pages, I show that these two methodologies, when used together, can allow us to build safety into a system or product right from the beginning of its development. Using more than a dozen practical examples from aerospace, nuclear, and other potentially hazardous facilities, I focus on methods for making logical decisions about high-consequence engineered systems and products in which safety is a key factor. In a nutshell, I demonstrate when, where, and how safety analysis fits into decision analysis, giving the necessary guidance on including safety in each key project decision.

I wrote this book for managers, project leaders, engineers, and scientists who create, design, develop, operate, or maintain complicated, high-consequence systems and products. It is also intended for students and anyone else interested in a broad perspective about the union of two technologies: decision analysis and PRA. I do not assume that the reader has expertise in either technology.

—Michael V. Frank

Acknowledgments

THE SUPPORT AND ENCOURAGEMENT of my family—my wife Jane and my children, Jeff, Heidi, and Heather—gave me the energy to start and complete this work. Karl N. Fleming, a longtime friend and professional colleague, was my first and best mentor in the technology of probabilistic risk assessment. I appreciate the continued moral support and enthusiasm of my father-in-law, Dr. Paul Griminger, professor emeritus at Rutgers University.

CHAPTER 1

Choosing Safety: An Overview

SIMPLY PUT, a decision is a choice among alternative courses of action. More complicated situations typically engender more difficult decisions because a decisionmaker (DM)[1] has many more interrelated factors to consider. Decision analysts (DAs) consider a good *decision* to be different from a good *outcome*. In the decision analysis context, a good decision has to do with *how* it is made, not with the final choice or outcome. According to Hammond and colleagues (1999), "The only way to learn to raise your odds of making good decisions is to learn to use a good decisionmaking process...."

In this book, I use examples to show how to make good decisions when system or product safety is involved. Such decisions do not necessarily seek maximum safety because the absolute maximization of either system or product safety may preclude other perspectives, such as overall cost or minimum needed capability. Instead, DMs seek a balance among all the factors.

The notion of maximizing safety implies that it can be measured or quantitatively analyzed. Probabilistic risk assessment (PRA), which had its genesis in the 1970s, provides the quantitative methodology that I use in this book for analyzing safety.

So what kinds of decisions do I analyze in these pages? I briefly describe some examples in the sections that follow.

1.1 A Pumping System Is Outdated

When new standards are promulgated, an older pumping system in a nuclear power plant must be replaced. Because it is used for cooling when the plant is shut down, the pumping system is important to safe plant operations. At least one of the three pumps in this system cannot rely on electricity for power because electric power to the plant is sometimes interrupted. Several different pumping systems, all of which are feasible, can meet the new standards. The systems vary in capital and operating cost, in the level of safety they bring to the plant, and in their availability for use during plant shutdown. How should the DM choose the pumping system?

Examples from Everyday Life

Making judgments about and relating disparate attributes, such as costs, product performance parameters, and safety to each other may seem odd. But, really, we make such decisions daily. For example, you wake up with a head cold during a flu epidemic. Do you go to work anyway or stay home to rest? Your company is in crisis and your co-workers could really use your help. On the other hand, you might become sicker if you go to work and you might infect others as well. Sound familiar? Whatever decision you make, you'll have weighed two attributes that are difficult to quantify—your career and your personal health.

Here's another common example. You've lived in your current location for many years with a satisfying lifestyle except for your employment. One day, you receive an ideal job offer, but it's in another state that doesn't have the type of lifestyle benefits you currently enjoy. Whatever you decide, you're once again weighing two difficult-to-quantify attributes: job satisfaction and lifestyle enjoyment.

1.2 A Ground Rover Needs Enhanced Reliability in Space

A spacecraft for interplanetary exploration is designed to deploy a small ground rover that moves along the planet's terrain, takes photographs, and analyzes rock samples. The rover must be able to wirelessly communicate data with the spacecraft lander. The lander relays the rover's data back to scientists on Earth, who use those data to plan the next day's rover movement. The scientists then transmit the commands back to the lander, which relays the instructions to the rover at the proper time. A successful mission depends on the reliability of the rover's communications system (a simple wireless modem) and on the rover's on-board electric power supply.

The team's scientists and engineers pepper the DM with suggestions for improving the rover, which range from using better software and enhancing maneuverability to supplying more electric power and increasing the rover's scientific capability. Each alternative differs in terms of its cost, its impact on the schedule, and the probability of achieving a successful mission. How should the DM decide which suggestions are best?

1.3 An Aircraft Door's Design Fails to Meet Standards

The design team manager of a new aircraft door operating system just found out that the current design will not meet government certification standards. Using PRA, the team develops a safety risk model for the door, and the results compare unfavorably to the government safety standards for aircraft certification. After some thought and discussion among the team members, four alternatives emerge: modify the existing design; start over, creating a new design; petition the government authorities to relax the standards; or continue the program without government certification. Each alternative differs in terms of its cost, its schedule for project completion, and the resulting level of airplane safety. How should the DM choose the course of action to follow?

1.4 A Wind Tunnel Experiment Could Be Dangerous

Engineers are modifying a wind tunnel to allow it to introduce pure oxygen in the model section (which contains the scale model of an aircraft), with a goal of studying the effects of air breathing in relation to the development of hypersonic aircraft. The wind tunnel burns a methane and air mixture to create a wind stream that flows through a nozzle at speeds ranging between Mach 4 and Mach 7. Methane and air mixtures can be explosive and can even detonate under some conditions, and, if something goes wrong, the wind tunnel configuration could create such conditions. Introducing oxygen into the wind stream would increase the probability and severity of a detonation. Different design options carry different levels of safety and cost. Should the project continue? If so, how should the DM choose the optimal design?

1.5 A Power Plant's Critical Equipment Could Flood

An audit of a power plant near a river found that one of its belowground rooms is open at ground level and can flood during a severe storm. If the room floods, the water will seep through the seals in the walls, which were installed to close off wall openings drilled to allow for passage of electrical cables and wires. The water will then flow into rooms that hold equipment critical to plant operation. In addition, equipment failure could lead to the release of hazardous gases.

The plant manager discusses the situation with company engineers and consultants and finds that this type of seal material has been known to degrade and eventually leak. He also learns that a newer type of seal material has much better long-term properties and is not so prone to leakage. These newer seals, however, are more expensive and more difficult to extract and replace if the electrical cabling must be replaced. The discussions result in five alternative courses of action: (1) seal off the room from the outside, (2) continuously monitor the current seals and repair or replace them in-kind as needed, (3) change all seals to the newer variety, (4) add flood-protection barriers around the critical equipment, and (5) do nothing. Although each alternative carries a different capital cost, operations cost, and safety level, plant capabilities are not affected by any of the alternatives. How should the DM decide which course of action to follow?

1.6 Some Definitions

All of the example decisions I give in this book involve complicated, high-consequence systems or products. *High-consequence* refers to a system or product whose failure can cause great harm, injury, or even death. A *complicated system* is difficult to analyze or understand, perhaps because it involves multiple interrelated factors or numerous internal and/or external interdependencies. Changes in such systems often give rise to difficulties in foreseeing consequences. For example, increasing the level of safety can coincide with increased overall cost and decreased product reliability. Ideally, a DM would like to know the ultimate outcome of selecting each alternative ahead of time. If the DM had a crystal ball or an oracle that could foresee alternative outcomes, selecting the best alternative would be easy. But because crystal balls and oracles are in short supply, the outcomes of any choice the DM makes are uncertain.

The term *decision attribute,* sometimes called *decision criterion* or *measure of effectiveness,* is a measurable or calculable factor used in deciding which alternative to choose. In the examples I present in this book, for example, the decision attributes are safety level, operating cost, capital cost, system availability, and mission success. *Multiattribute decisions* involve more than one attribute, and each alternative has an outcome associated with each attribute. *Decision analysis* is the field that was developed to help guide DMs through a cogent, rational method for choosing among alternative courses of action. Following this method increases a DM's understanding of the interrelationships of attributes and alternatives. Although it does take time to go through the process, making potentially dangerous and expensive systems safer demands that this time be taken.

Throughout this book, I also use the following definitions:

- *Risk*—the probability of failure, harm, loss, damage, injury, or other undesirable event
- *Probabilistic risk assessment*—a method that quantifies safety risk
- *Risk management*—a process of using risk assessment to make decisions about maintaining a desired level of risk
- *Decisionmaking*—a process of choosing among alternative courses of action
- *Decision analysis*—a logical method that aids in decisionmaking
- *Reliability*—the probability that an object can perform its intended function for a specified interval under stated conditions and

- *Reliability analysis*—a method that quantifies reliability.

1.6.1 A New Way of Thinking: Safety as a Number

It has long been common practice among those who manage engineering projects to calculate cost and schedule consequences of various alternatives. Calculating safety as a *number*, though, is relatively new and is not yet in common practice. In this book, I emphasize that safety is *quantifiable* using PRA (sometimes called probabilistic safety assessment [PSA]).

PRA (the term I use in this book) is a method for calculating the level of safety by numerically estimating risk. In this way, the level of safety can be quantitatively included in a decisionmaking process with the same rigor as other quantified attributes such as cost, schedule, reliability, and many more. Calculating safety as a number, or more commonly, as a *probability distribution,* allows DMs to treat safety as they would other attributes in a multiattribute decision process.

Figure 1-1 is a simplified graphic of the concept of safety-related decisionmaking. At the left of the figure is the point in time when a decision must be made (labeled "Choose among alternatives"). Project engineers or scientists develop alternative courses of actions (labeled A, B, and N), and consult with the DM (e.g., the project manager) to create a set of attributes. The figure shows three typical attributes—cost, performance, and safety.

Next, DAs work with the engineers and scientists to estimate the probable consequences (or *effects*) of each alternative with respect to each attribute.[2] The effects in Figure 1-1 are numerical quantities (the probability distributions), which yield a metric of each attribute. Multiattribute decision analysis combines all the attribute consequences with the DM's values or preferences to arrive at a ranking of alternatives.

The DM then decides which set of effects best meets the project's objectives. As I describe in detail in this book, choosing the best set of effects involves combining *consequences* with *values*. As expressed by Keeney (1992), "Values are principles used for evaluation. We use them to evaluate the actual or potential consequences...." Said another way, values are a basis for establishing preferences. Without personal or organizational principles, we have no criteria with which to evaluate attributes and make the best choices. The DM infuses the decision analysis with her own values, allowing her to develop preferences among the attributes. For example, for purposes of the project's objectives, she may value safety as much more important than cost or performance. Often the attributes must be considered jointly. For example, the incremental cost of additional performance and/or additional safety may be an important consideration.

In this context, we can think of performance as a desired system or product design feature. Examples include weight-carrying capability for rockets, power output for generators and microwave ovens, acceleration for automobiles, and durability for children's toys. And we can think of safety as the probability of

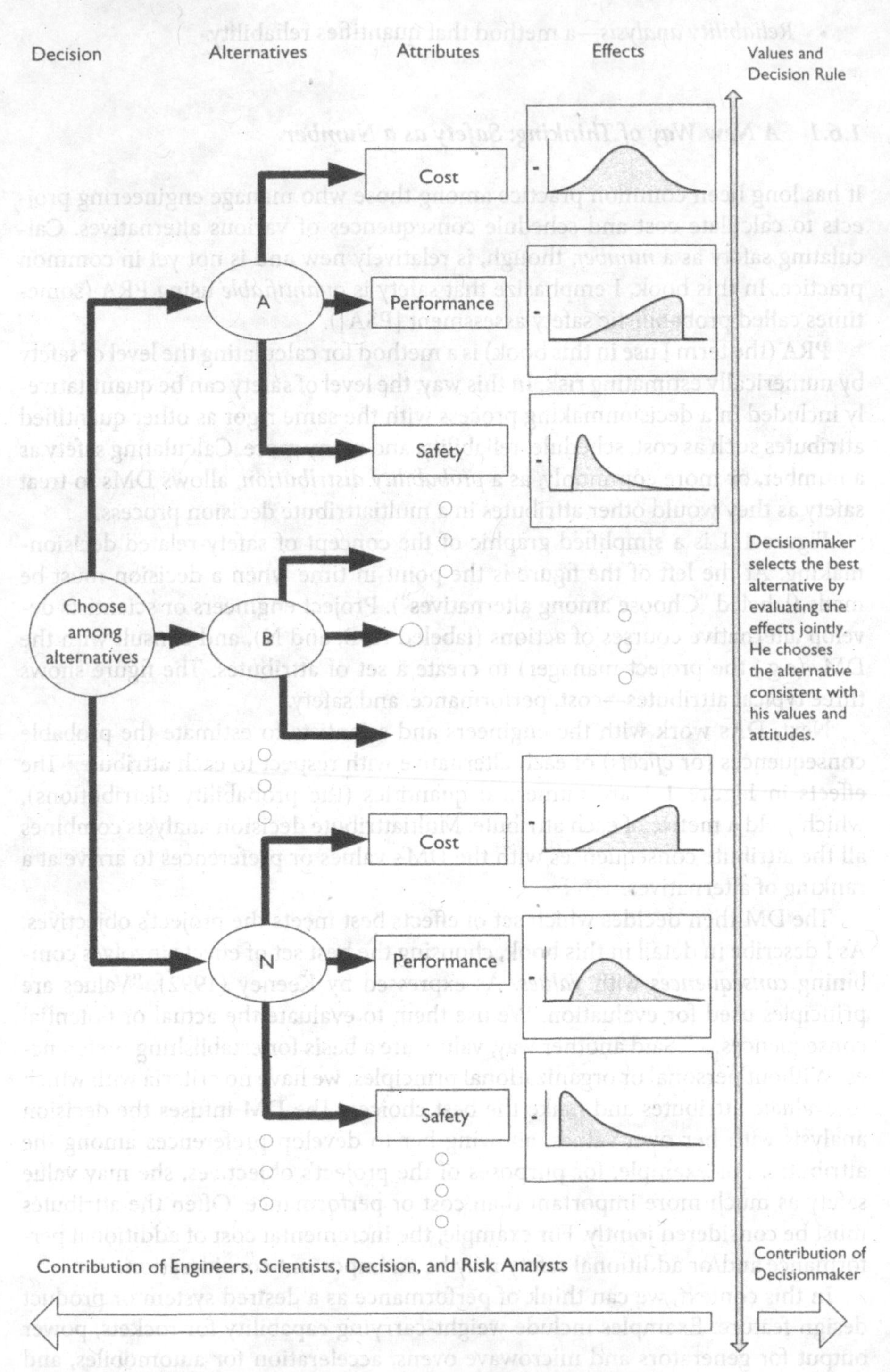

Figure 1-1. ***Overview of Safety-Related Decisionmaking***

being free from harm. Finally, we can think of cost in terms of metrics such as investment expenditures or loss expressed in dollars.

1.6.2 The Risk Management Connection

Risk management is a procedure in which we attempt to manage the future attributes of a product (such as cost, performance, and safety) by frequently reevaluating these attributes to determine if the project is proceeding satisfactorily. If there is an unsatisfactory aspect, we develop corrective alternatives and analyze each one to select the best. In other words, a decision analysis process is used periodically during the project. Such an iterative evaluation process of controlling risk while a system is under development or during operation, as conceptually illustrated in Figure 1-2, is an instance of risk management.

Risk goals can be associated with variables such as safety, reliability, cost, and schedule, among others. *Risk models* (e.g., safety, reliability, and cost models) are

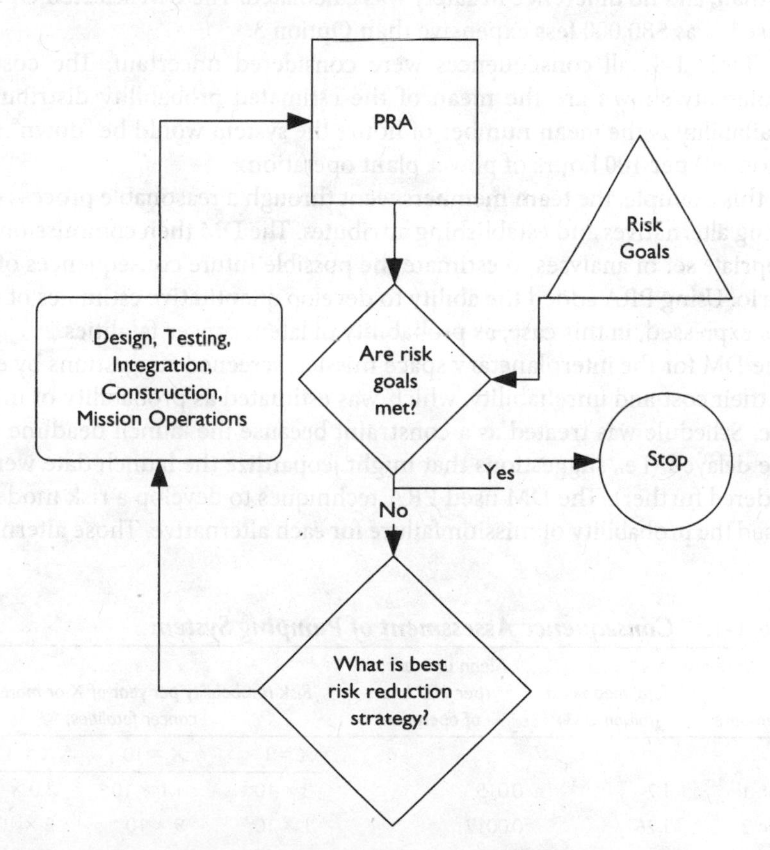

Figure 1-2. *Flow Diagram of a Risk Management Process*

Note: PRA = probabilistic risk assessment.

developed and evaluated at prescribed intervals during system design and operation. At any point that a goal is not met, alternatives to reduce risk are developed. To select the best risk reduction strategy, a decision analysis is conducted, using the process illustrated in Figure 1-1.

1.7 The Example Decisions, Explored

The nuclear power plant's DM paid for analyses of three attributes: total cost, system unavailability, and safety (see Table 1-1). Of interest for safety was a risk metric expressed as the probability of various levels of latent cancer fatalities associated with accidents that are mitigated by this pumping system. Option 1 was eliminated—even though it was less costly—because the DM believed that it carried significantly higher unavailability and safety risk. The DM considered the difference between Options 2 and 3 (with respect to unavailability) to be very small, and no difference in safety was calculated. The DM selected Option 2 because it was $80,000 less expensive than Option 3.

In Table 1-1, all consequences were considered uncertain. The cost and unavailability shown are the mean of the estimated probability distributions. Unavailability is the mean number of hours the system would be "down" (non-operational) per 100 hours of power plant operation.

In this example, the team members went through a reasonable process of developing alternatives and establishing attributes. The DM then commissioned an appropriate set of analyses to estimate the possible future consequences of each scenario. Using PRA added the ability to develop quantitative estimates of safety risk as expressed, in this case, as probability of latent cancer fatalities.

The DM for the interplanetary space mission screened suggestions by evaluating their cost and unreliability, which was estimated as probability of mission failure. Schedule was treated as a constraint because the launch deadline could not be delayed (i.e., suggestions that might jeopardize the launch date were not considered further). The DM used PRA techniques to develop a risk model that assessed the probability of mission failure for each alternative. Those alternatives

Table 1-1. ***Consequence Assessment of Pumping System***

Alternative	*Total mean cost (million US$)*	*Mean unavailability (per 100 hours of operation)*	*Risk (probability per year of X or more latent cancer fatalities)*		
			X = 1	X = 10	X = 100
Option 1	1.2	0.015	3×10^{-5}	1.1×10^{-5}	1.0×10^{-5}
Option 2	1.26	0.0012	1×10^{-5}	9×10^{-6}	8×10^{-6}
Option 3	1.34	0.0011	1×10^{-5}	9×10^{-6}	8×10^{-6}

Source: Frank et al. 1981.

that decreased the probability of a mission failure and decreased cost were implemented before all others. Those that increased probability of mission failure or had indeterminate effects were eliminated from consideration. Ultimately, the DM opted to reorient resources to concentrate on high communications reliability and deemphasize software development. This DM understood the risk management principle that frequent reevaluation of attributes when compared to goals helps to determine if the project is proceeding satisfactorily.

The decision about how to continue the development of the aircraft door was made relatively easily. Relaxing the government's requirements was not possible, and top project management insisted that the airplane be certified. This left the alternatives of starting over and modifying the original design. Because both alternatives had to meet the safety requirements, it became a question of which alternative would be less expensive and offer the best chance of meeting the schedule for the airplane's first flight. Modification of the current design was fraught with uncertainties about cost and schedule because it was being pursued by a group that had little prior experience in aircraft design and manufacture or with federal government regulations. The DM found an experienced vendor who suggested a new design with much less uncertainty in terms of cost and schedule. As the new design progressed, the safety risk model was modified to demonstrate compliance with the government safety standards.

This example highlights two important aspects of decisionmaking. First, uncertainties are often important in making a decision. Because probabilities are one form of mathematical expression of uncertainty, risk goals may be formulated as, for example, probability of exceeding the budget by an amount greater than *x*, probability of exceeding the schedule by an amount greater than *y*, and probability of mission failure. Whenever an estimated probability of one of the important decision attributes exceeds the goal, a risk reduction loop should be completed (refer to Figure 1-2). In this example, the DM judged that the uncertainty (i.e., risk) of continuing with the current design was larger than the risk of starting over with an experienced vendor.

Second, overall objectives or goals should be clearly formulated and articulated. Until management clearly set forth the policy that aircraft certification must be achieved, goals were not articulated in this example. It is much easier to evaluate whether a project is proceeding satisfactorily if we can evaluate progress against a well-defined objective. Similarly, it is much easier to develop meaningful attributes against which to measure alternatives if the objective is clearly defined. Finally, it is also much easier to judge which alternative is best, as measured by the calculated effects, if we can assess them in relation to a clear objective.

The mission of the wind tunnel, helping to develop hypersonic aircraft, was judged important enough to continue with the modifications. Management also judged that the level of safety risk before the modifications was acceptable. The goal, then, was to proceed with the introduction of an oxygen system and make other modifications that would reduce the safety risk in an attempt to compensate for the increased risk of introducing oxygen. A PRA performed on the wind tunnel suggested ways in which to reduce the risk of the wind tunnel and generated a number of viable alternatives. When comparing strategies of comparable risk reduction and estimating cost on the basis of initial capital outlay, the DM selected the least expensive approaches.

This decisionmaking process was based solely on initial capital outlay and safety risk. One of the things that the DM might have done better would have been to develop complete life-cycle costs that include—as a minimum—the costs of the following: operations, maintenance, downtime, spare-part procurement, and extra staff. Additionally, a broad articulation of a project objective might have changed or added attributes. For example, convenience of use to customers, availability of the overall wind tunnel, turnaround time between tests, and overall life-cycle costs of the tunnel might have been included in any decisionmaking about alternative modifications.

Management of the power plant near the river paid for a study that estimated the probability of a flood severe enough to cause leaks through the seals. The study, which essentially assessed the viability of doing nothing, found that the probability of such a flood was unacceptably high.

Management then commissioned a cost study, which found that the replacement of the seals was the least expensive alternative. Failure to replace seals would mean either having to protect most of the plant components or making the building itself waterproof. Both alternatives were extremely costly and would interfere with plant operations. Management was reluctant to install a new seal material without sufficient assurance that it would perform well in the long term. A series of tests in which the new material successfully performed when subjected to extreme environments convinced management to implement this alternative.

The DM sought new information even after the attributes and alternatives had originally been defined. This is good practice and certainly helped to reach a reasonable conclusion. One of the things that the DM might have done better, however, was to think about new alternatives as new information was being gathered. For example, management might have attempted to discover the particular environmental effect (if any) that led to the degradation of the old seals. Then, new alternatives that included modifying the environment or finding another alternative seal material might have been considered.

1.8 Formalizing the Process: Decision Analysis

All of these decisionmaking processes were logical and reasonable, although ad hoc in nature. The DMs made their selections on the basis of minimizing risk. Each of the decisions in the examples was made by an individual. The decision, in each case, was influenced by the DM's implicit or explicit preferences or attitudes. For example, the power plant DM implicitly expressed an opinion that the difference in safety risk and availability between the do-nothing alternative and the other alternatives was sufficiently large to pursue the other alternatives. This DM put a high value on safety. Or, in other words, (1) safety was preferred

a lack of knowledge about it, or both. One aspect missing from this dictionary definition is the notion of the *severity* of the danger or the *amount* of loss. The concept of amount of loss, danger, harm, or injury of an event is as fundamental to the definition of risk as the probability of the event. In a simple example, a 1% chance of losing $10 has a lower risk than a 1% chance of losing $100. One of the early definitions of risk used in probabilistic risk assessments (PRA) is called expected loss, $\overline{L}$, and is expressed as

$$\overline{L} = \int p(l) l dl \tag{2-1}$$

where $p(l)$ is the probability of loss or harm. We can see, then, that both the probability of the level of loss and the amount of loss, l, are essential aspects of risk.

Another definition emerged from applying PRA in the nuclear power industry. Kaplan and Garrick (1981) give a useful definition associated with estimating risk as a composite of separate scenarios. To arrive at their definition, they asked three questions:

1. What can go wrong?
2. How likely is it to happen?
3. What are the consequences?

The first question is answered by defining accident scenarios, which are sequences of events that lead to undesirable outcomes. The answer to the second question is the probability of those sequences occurring. Each scenario leads to an end state, which defines the consequences and answers the third question. In the original article by Kaplan and Garrick, risk is defined as a "triplet," which is a set composed of scenario, probability, and consequence. Many practitioners, though, quantify risk as a summation of the product of consequence and probability of each scenario. This is also, in effect, an expected consequence as follows:

$$\overline{C} = \sum_{i=1}^{n} c_i p_i \tag{2-2}$$

Here, i is the ith of n scenarios and c_i and p_i are the consequence and probability associated with the ith scenario, respectively.

Risk, however, need not be represented as an expected consequence. A more general representation is a risk curve, which is a probability distribution that represents probability of consequence. In this representation, the expected consequence is merely the mean of the distribution, as illustrated in Figure 2-1.

The curve labeled CDF is a cumulative distribution function. This is read as the probability along the y-axis of the consequences being *less than* or equal to the quantity[3] on the x-axis. For example, there is approximately a 38% chance (i.e., probability) that the consequences will be less than or equal to the mean of 2. The CCDF is a complementary cumulative distribution function and its quantities are the probabilistic complement (i.e., 1 minus) of the CDF. This is read as the probability along the y-axis of the consequences being *greater than* or equal

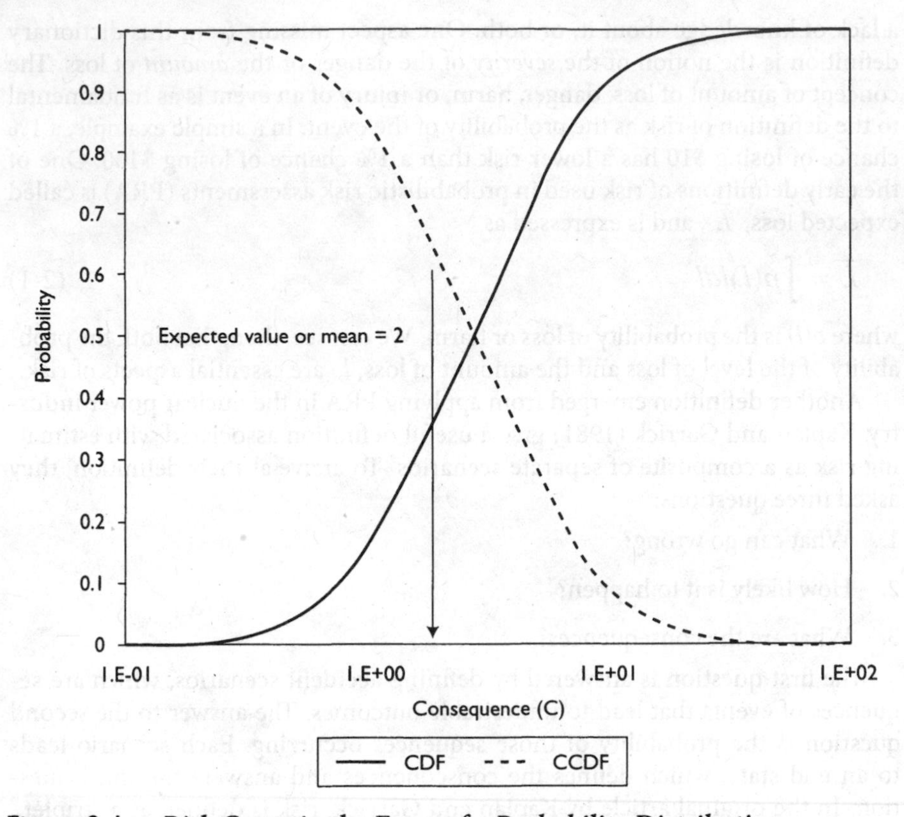

Figure 2-1. ***Risk Curve in the Form of a Probability Distribution***

Notes: CDF = cumulative distribution function; CCDF = complementary cumulative distribution function.

to the quantity on the x-axis. For example, there is approximately a 62% probability that the consequences will be greater than or equal to the mean.

In the Kaplan–Garrick formulation, the probability of each scenario can be represented as a curve such as that shown in Figure 2-1. In a simple binary application, each scenario would end in one of two end states—*OK* or *Loss.* The total probability of the end state, Loss, would be the probabilistic sum of all individual scenarios that end in Loss. In this context, risk would be defined as the probability of Loss.

2.1 The Difference between Hazard and Risk: The Feyzin Disaster

In this section, I summarize the sequence of events of an actual disaster to illustrate the difference between hazard and risk (Kletz 1994). Liquid gas tank farms are used to store flammable liquids such as liquid propane gas (LPG), liquid natural gas (LNG), and gasoline. At a liquid flammable gas tank farm in Feyzin, France, LPG was stored in spherical tanks, and gasoline and other fuel oil in cylindrical tanks. The underlying hazards of stored flammable materials, should the tanks leak or discharge their contents in an uncontrolled manner, would be environmental pollution, fire, and explosion. Should the hazards be actualized, possible consequences would be illness from inhalation of hazardous gases, property damage, and injury or death from fires and explosions. At Feyzin, the following combination of events, called a *scenario*, allowed the hazards to actualize and lead to fire and explosions on January 4, 1966.

Because the bottoms of storage tanks collect water, the maintenance crew began to perform a routine draining of a spherical LPG tank, opening drain valves A and B (shown in Figure 2-2). When traces of propane appeared in the effluent, the workers closed valve A and then opened it a little to complete the draining. But a clog had developed, obstructing flow, so the crew opened valve A fully, which allowed the clog to clear. Liquid propane gushed out.

The handle of valve A broke off and could not be replaced, and valve B froze or stuck in the open position and could not be moved. Propane continued to flow from the tank, and because it is heavier than air, a visible cloud 1-m deep spread to a distance of 150 m. Workers called the police, who barred entry to the main road fronting the tank farm. But a car was idling on a side road, and a spark from its engine ignited the propane, which had spread out enough so that its mixture with air was within the limits of flammability.

The fire flashed back to the tank and began burning the drained propane, which continued to drain from the tank. As shown in Figure 2-3, the spherical

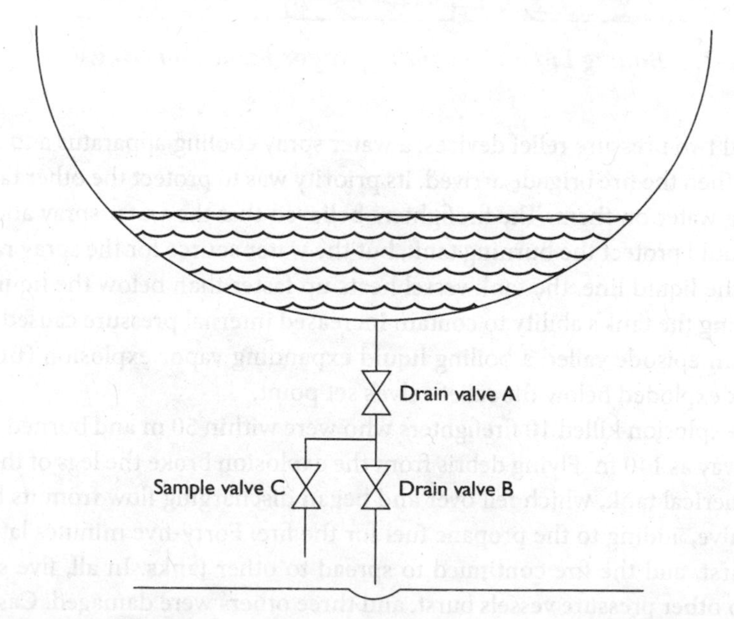

Figure 2-2. ***Drain System of Spherical Liquid Propane Gas (LPG) Tank***

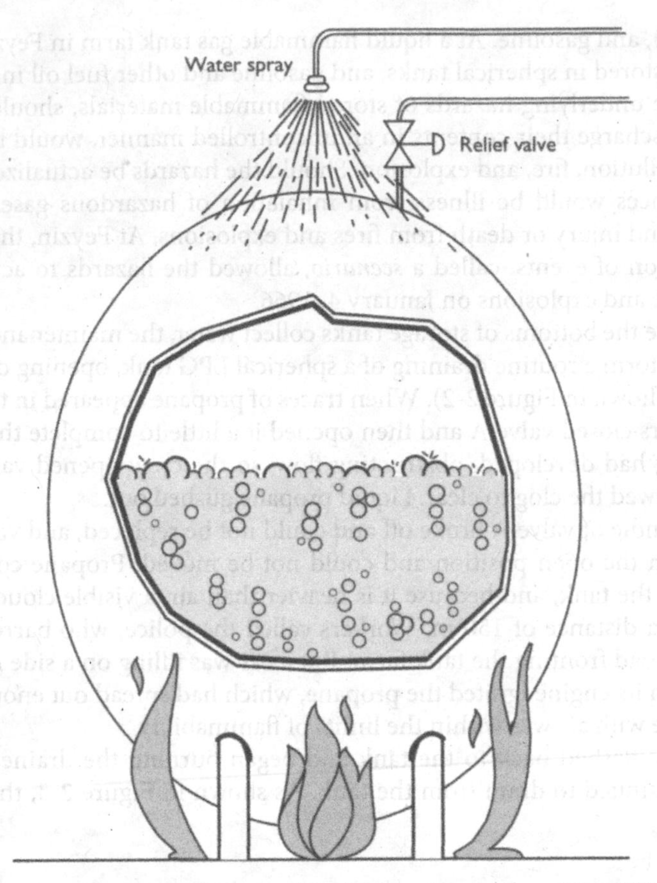

Figure 2-3. ***Boiling Liquid Expanding Vapor Explosion (BLEVE)***

tank had two pressure relief devices: a water spray cooling apparatus and a relief valve. When the fire brigade arrived, its priority was to protect the other tanks by spraying water on them. The firefighters believed that the water spray and relief valve would protect the burning tank, but the water source for the spray ran dry. Above the liquid line, the tank vessel heats up faster than below the liquid line, weakening the tank's ability to contain increased internal pressure caused by the fire. In an episode called a boiling liquid expanding vapor explosion (BLEVE), the tank exploded below the relief valve's set point.

The explosion killed 10 firefighters who were within 50 m and burned people as far away as 140 m. Flying debris from the explosion broke the legs of the adjacent spherical tank, which fell over and began discharging flow from its broken relief valve, adding to the propane fuel for the fire. Forty-five minutes later, this tank burst, and the fire continued to spread to other tanks. In all, five spheres and two other pressure vessels burst, and three others were damaged. Casualties included 15 to 18 deaths and 80 injuries (Kletz 1994).

This example relates to the Kaplan–Garrick (1981) definition of risk as I describe here. The first part of the triplet is the scenario. The scenario was initiated by a routine maintenance action and a series of subsequent valve failures and inaccurate assumptions by the fire brigade, which led to the fire and explosions. Notice that the actualization of a hazard may be part of a scenario. The last part of the triplet is the consequences (injury, death, and property damage).

The middle part of the triplet is the probability of occurrence of a scenario leading to the consequences. In the next section and Chapter 3, I discuss the concept of probabilities and methods for estimating them during a PRA.

Probabilities must be viewed within the context of many potential alternative scenarios. Probability estimation has meaning only in a prospective context, not a retrospective context.

Before the Feyzin accident, the operators knew that fires and explosions, injuries, and deaths were possible, but they did not understand the probability of such events. In other words, they were uncertain about what might happen. If the operators had asked, "What is the risk of a fatal accident here?" probability estimation would have been a useful undertaking. If a PRA had been performed, the operators might have gained an understanding about the likelihood of occurrence (or odds against occurrence) of the various events (such as valve failures).

After the disaster, the perspective changes. One question that is often asked after the fact, usually with some dismay, is something like "What are the odds?" Here the questioner is not looking to the future and asking about what might happen because the particular accident has already happened. There is no uncertainty, and the probability of that particular accident having already occurred is 1. A more useful question at this time would be, "Given that this accident has occurred, what would be the probability of it happening again here or in some other similar facility?" Probability estimation within the context of a PRA would be useful in answering this question. The numerical estimation of a probability, therefore, depends on the questions being asked and the timing of these questions.

2.2 Uncertainty and Probability

Probabilities are a mathematical measure of uncertainty. By convention, probabilities range from zero (no chance of happening) to one (certain to happen). The local weatherman assigns a chance to his prediction of rain because he doesn't know for sure if it will rain. In the north coastal region of San Diego County, it does not rain about 350 days a year. Weathermen get excited when there is rain possible, and they give a probability such as 20% chance of rain tomorrow morning increasing to 50% by late afternoon.

In another example, Weather.com gives hour-by-hour probabilities to account for the constantly varying atmospheric conditions that might cause rain.

Variability in such natural phenomena is sometimes called *aleatory uncertainty* (Apostolakis 1994). Decision and risk analysts assign a probability to future events because of these types of uncertainty.

Sometimes it doesn't rain at all and sometimes it rains when not predicted. The weathermen, therefore, don't precisely know the probability of rain. The probability that they give is uncertain. Meteorologists use computer models to make predictions about the weather. However, these models are only approximate. They do not fully account for all of the factors, effects, and interrelationships that lead to tomorrow's weather. Furthermore, the inputs to these computer models rely on weather stations that are widely scattered and, therefore, represent only an approximate measure of the atmospheric conditions such as temperature and pressure. Uncertainty that stems from incomplete or incorrect knowledge is often called *epistemic uncertainty*. The weatherman should really give a range of rain probabilities to account for this kind of uncertainty.

Similarly, we live in a dynamic world where conditions are always changing. The information about future events (such as future costs) used by decision and risk analysts is incomplete, and the models used are only approximations. Estimates of risk, then, also have epistemic and aleatory uncertainties. When the probabilities themselves are uncertain because they vary with the severity of the effect, because they result from changing conditions, or because there is insufficient knowledge to calculate them precisely, analysts use *probability distributions* to represent uncertainty.

What is a probability distribution? It's a way of representing uncertainty. Table 2-1 shows a convenient way to begin thinking about probability distributions—as a set of ordered pairs of probabilities and attributes (such as cost).

In this example, the analysts don't precisely know the development cost of the item (such as the airplane door in Chapter 1), but they think that it's somewhere between $10 and $40 million. Their best estimate is that there is a 50% chance the cost will be about $30 million, with smaller chances of the cost equaling $10, $20, or $40 million. Figure 2-4 is a graphical representation of Table 2-1. In the figure, the total of all discretely defined probabilities must total 1 or 100%.

Table 2-2 shows a general tabular form of a probability distribution. Note that in tabular form the distribution is discrete (as opposed to continuous) in that it is defined only at specific points.

Table 2-1. ***Example Tabular Form of Probability Distribution***

Probability (%)	*Cost (million US$)*
10	10
30	20
50	30
10	40
Total = 100	

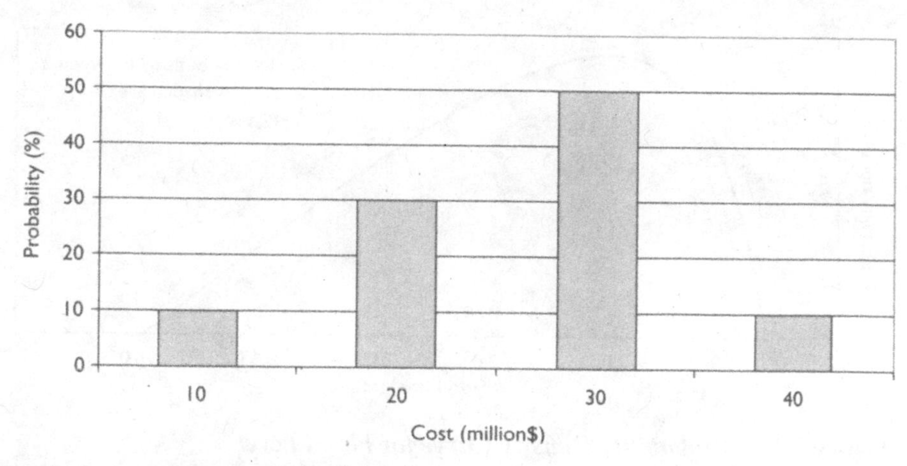

Figure 2-4. *Discrete Probability Distribution*

The elements in the right-hand column of Table 2-2 can be any uncertain, quantifiable attribute such as cost, time to complete a task, height of flood water, safety risk, or reliability, among others.

Some variables are better exhibited by a continuous curve rather than a discrete set of tabular or graphical numbers. For example, the level of flood waters at the plant near the river (from Chapter 1) is a continuous function. The likelihood of occurrence of floods may be expressed as a frequency (number of occurrences per year). There are two common ways to graphically exhibit such continuous probability distributions. One is called a probability density function (PDF). The area under a PDF equals 1, and probabilities are defined as slices of this area as shown in Figure 2-5. For a continuous curve, we can speak of a probability only over an interval. For example, the curve shows that the probability of a flood between 2 and 3 m is 0.48 because this is the area under the curve. A

Table 2-2. *General Tabular Form of Discrete Probability Distribution*

Probability	*Attribute*
P_1	A_1
P_2	A_2
P_3	A_3
—	—
—	
P_i	A_i
—	—
—	—
P_N	A_N
Total = 1	

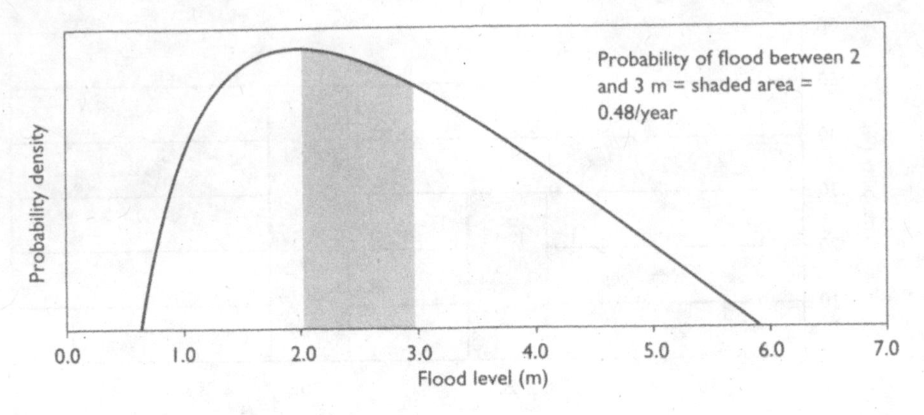

Figure 2-5. *Probability Density Curve for Flood Level*

PDF is the derivative of either a CDF or CCDF, as I discussed previously. If the PDF in Figure 2-5 is integrated, a curve like that shown in Figure 2-6 results. Such a curve is often preferred because probabilities can be read directly from the curve. Each point on the CCDF curve in Figure 2-6 shows the probability of being greater than or equal to the corresponding flood level. For example, the probability of a flood greater than or equal to 4 m = 0.025 and the probability of a flood between 2 and 3 m is read as the difference between the y-axis frequencies at 2 and 3 m, respectively.

Risk analysts often use the probability of an undesired consequence as a measure of risk. For example, the analysis of the wind tunnel (Chapter 1) used the "probability of an explosion over 30 years" as the risk metric, and the analysis of the spacecraft used "probability of mission failure" as a metric. Because the analysis is trying to evaluate events that might or might not happen, using probability as a risk metric is appropriate. Because the models and inputs are approximate, it is also appropriate that these probabilities are, themselves, uncertain and represented by probability distributions.

2.3 Risk and Uncertainty

Suppose that we're all omnipotent and omniscient beings. We know everything that has and will occur in the macroscopic (nonquantum) world. We can do whatever it takes to change events we'd like to avoid. If this were true, would there be a risk?

The answer is no. If we were all-knowing and present everywhere all the time, we'd be able to see all possible outcomes of all our possible decisions and actions. Risk arises from our inability to reliably predict the future. Decisions we make and actions we take involve risk because we don't know what will happen when we take them. Another way of saying this is that the future is uncertain and, as

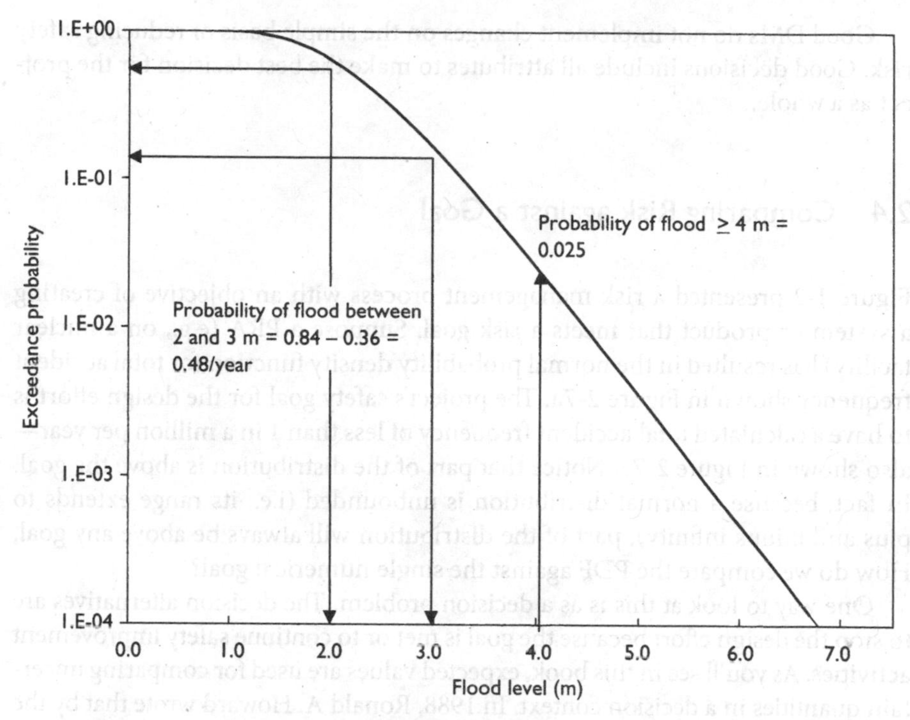

Figure 2-6. ***Complementary Cumulative Distribution Function (CCDF) Curve for Flood Level***

a result, any actions we take involve some risk. The need for risk assessment, therefore, arises from such uncertainties. The notion that risk is inexorably tied to uncertainty is fundamental to understanding PRA and decision analysis.

Decisions about safety are important in high-consequence systems, which are those having a large potential for injury, property damage, or expensive system downtime. Decisions that affect safety are made from the beginning of a project through its end. Any concept, design, development effort, operation, maintenance plan, or staffing action—along with a myriad of other project aspects of high-consequence systems—affects safety. Decisions involving safety are rarely one-dimensional. A decisionmaker (DM) will want answers to questions such as

- What are all of my options?
- What will they cost?
- What is the current risk and how much will each option change it?
- How will each option change reliability and performance?
- How much will implementing each option increase schedule to complete the project?

Good DMs do not implement changes on the simple basis of reducing safety risk. Good decisions include all attributes to make the best decision for the project as a whole.

2.4 Comparing Risk against a Goal

Figure 1-2 presented a risk management process with an objective of creating a system or product that meets a risk goal. Suppose a PRA (e.g., on a nuclear facility) has resulted in the normal probability density function for total accident frequency shown in Figure 2-7a. The project's safety goal for the design effort is to have a calculated total accident frequency of less than 1 in a million per year—also shown in Figure 2-7a. Notice that part of the distribution is above the goal. In fact, because a normal distribution is unbounded (i.e., its range extends to plus and minus infinity), part of the distribution will always be above any goal. How do we compare the PDF against the single numerical goal?

One way to look at this is as a decision problem. The decision alternatives are to stop the design effort because the goal is met or to continue safety improvement activities. As you'll see in this book, expected values are used for comparing uncertain quantities in a decision context. In 1988, Ronald A. Howard wrote that by the rules of probability, the mean of a PDF of an uncertain event is the probability that must be assigned to the occurrence of that event. His article proved that the mean value is the appropriate substitute for a probability distribution when making an immediate decision (Howard 1988). In this context the mean value of a frequency, $\bar{\lambda}$, is the first moment of the underlying PDF, $p(\lambda)$, where the mean is defined as

$$\bar{\lambda} = \int \lambda p(\lambda) d\lambda$$

The U.S. Nuclear Regulatory Commission (NRC) has stated, "Commission has adopted the use of mean estimates for purposes of implementing the objectives of this safety goal policy…." (NRC 1988).[4]

In this approach, if the mean value of the underlying probability distribution of accident sequence frequency is less than the goal, the goal is met.

Another way to make the comparison, however, is to consider the amount of the distribution below the goal. We can do this with a statement that indicates the probability of meeting the goal. For example, a DM may be more comfortable by creating a goal that gives X% probability (e.g., 95% probability) of meeting the goal. The mathematical interpretation of such a statement would be

$$p_G = \int_{-\infty}^{G} p(\lambda) d\lambda$$

where p_G = the probability that the goal, G (e.g., 1E-06/year), is met. The integral is the area under the PDF in Figure 2-7a that is below the goal. In this way, a DM can decide on the level of assurance (e.g., X%) that the system is safe. In this example, if p_G is at least 95%, the design effort has met the goal.

Suppose a group of engineers were assigned to design and analyze a new type of container. Because the container is to hold a hazardous material, it must not release its contents. During its lifetime it may be exposed to a variety of hazards, such as fires and earthquakes. Because it is not feasible to design to the worst conceivable fire or earthquake, the engineers must make a decision about the robustness of the vessel against the hazards. In decision terminology, a choice must be made about the acceptable level of risk associated with exposure to hazards. Earthquakes and fires come in all sizes (i.e., severity) and each size has a probability of occurrence. The engineers characterize the range of future fires or

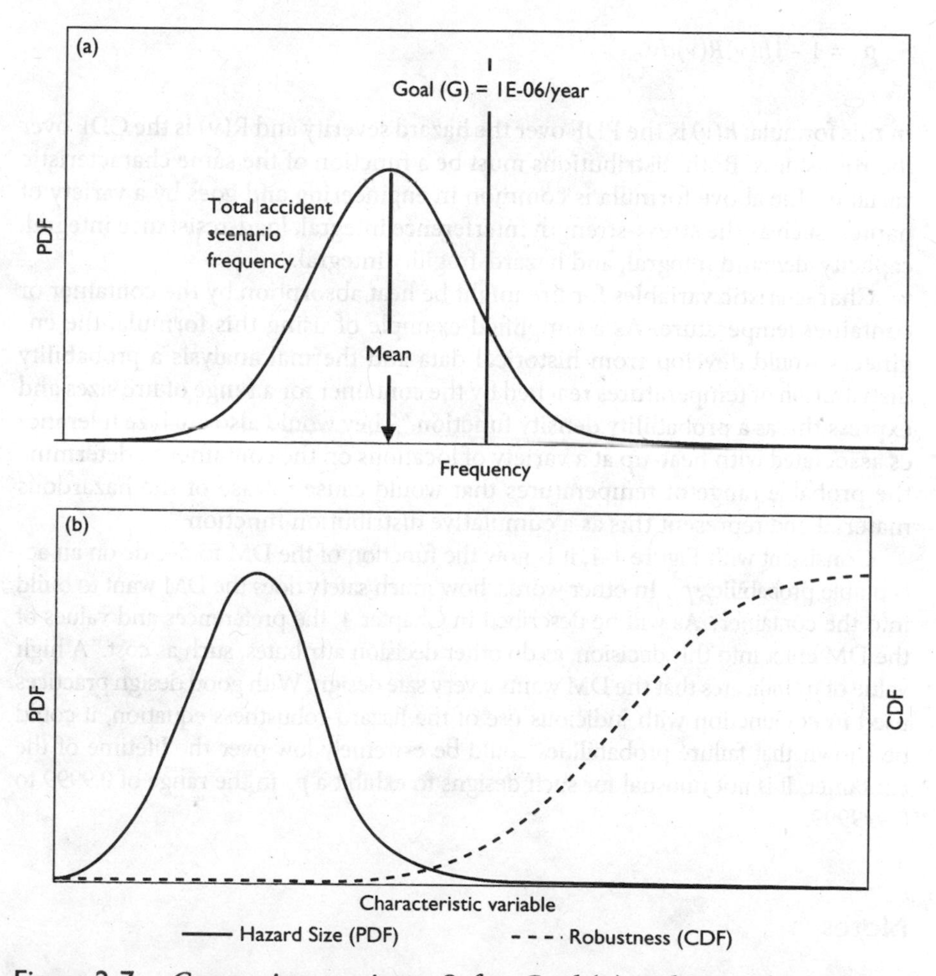

Figure 2-7. *Comparison against a Safety Goal (a) and Hazard Size and Robustness of the Comparison (b)*

Notes: PDF = probability density function; CDF = cumulative distribution function.

earthquakes that might occur with a PDF as shown in the left-hand curve of Figure 2-7b.[5] The robustness of the vessel is also uncertain because of such factors as variations in manufacturing tolerances, material properties, and normal wear and tear during use. The characterization of uncertain robustness is shown in the right-hand curve of Figure 2-7b. If we compare a and b in the figure, we see that it is a similar decision problem with the addition of uncertainty in the goal. If both the hazard and robustness (or goal) are represented by their respective mean values, the design might be considered robust enough if the mean value of the hazard severity is less than the mean value of robustness.

A much better way to make the decision, however, is to obtain the mean probability that the hazard severity will be less than the robustness. In this case, p_G is derived by integrating over the range of the characteristic variable, ν, as follows:

$$p_G = 1 - \int h(v)R(v)dv$$

In this formula, $h(v)$ is the PDF over the hazard severity and R(ν) is the CDF over the robustness. Both distributions must be a function of the same characteristic variable. The above formula is common in engineering and goes by a variety of names, such as the stress-strength interference integral, load-resistance integral, capacity-demand integral, and hazard-fragility integral.

Characteristic variables for fire might be heat absorption by the container or container temperature. As a simplified example of using this formula, the engineers would develop from historical data and thermal analysis a probability distribution of temperatures reached by the container for a range of fire sizes and express this as a probability density function.[6] They would also analyze tolerances associated with heat-up at a variety of locations on the container to determine the probable range of temperatures that would cause release of the hazardous material and represent this as a cumulative distribution function.

Consistent with Figure 1-1, it is now the function of the DM to decide on an acceptable probability, p_G. In other words, how much safety does the DM want to build into the container? As will be described in Chapter 4, the preferences and values of the DM enter into this decision, as do other decision attributes, such as cost. A high value of p_G indicates that the DM wants a very safe design. With good design practices used in conjunction with judicious use of the hazard-robustness equation, it could be shown that failure probabilities could be extremely low over the lifetime of the container. It is not unusual for such designs to exhibit a p_G in the range of 0.9999 to 0.999999.

Notes

1. *Merriam-Webster Collegiate Dictionary*. Unabridged, online. 11th ed. s.v. "Safe." http://unabridged.merriam-webster.com/cgi-bin/collegiate?va=safe&x=0&y=0 (accessed May 9, 2007). Definitions summarized here.

2. *Webster's New World Dictionary of the American Language.* 1962. College Edition. New York: The World Publishing Company. Definitions summarized here.

3. A common usage of the word *value* relates to the magnitude, amount, or quantity of an independent variable of a function as in $f(X=x)$, where x is the value of the independent variable X. In this book, however, I use the term value in its decision theory sense of relating to a decisionmaker's preferences. I don't change the terms *expected value* and *expected utility* from their common usage in decision theory as a probability-weighted average.

4. See Chapter 10 for more on the NRC safety goal policy.

5. Of course, any hazard may be characterized by a probability distribution with respect to size or severity in this way.

6. Analysis of the effect of a fire is far more complicated, but a complete description is not in the scope of this book. I want to simply describe the use of the formula for decisionmaking.

References

Apostolakis, G. 1994. A Commentary on Model Uncertainty. In *Proceedings of Workshop I in Advanced Topics in Risk and Reliability Analysis. Model Uncertainty: Its Characterization and Quantification*, edited by A. Mosleh, N. Siu, C. Smidts, and C. Lui. NUREG/CP-0138. Washington, DC: U.S. Nuclear Regulatory Commission (NRC), 13–22.

Howard, R. A. 1988. Uncertainty about Probability: A Decision Analysis Perspective. *Risk Analysis* 8(1).

Kaplan, S., and B.J. Garrick. 1981. On the Quantitative Definition of Risk. *Risk Assessment* 1(1): 11–27.

Kletz, T. 1994. *What Went Wrong: Case Histories of Process Plant Disasters*, 2nd ed. Houston, TX: Gulf Publishing.

NRC. 1988. *Safety Goals for the Operations of Nuclear Power Plants; Policy Statement.* Title 10, *U.S. Code of Federal Regulations,* Part 50, November 30.

CHAPTER 3

Probabilistic Risk Assessment (PRA) Concepts and Methods

EVERY DAY, PROJECT AND PROGRAM MANAGERS grapple with questions about the risks of their projects: What's the risk of this mission? What's the risk of core damage? Is it safe to launch? What's the risk of a toxic gas leak? What's the risk of budget overrun? What's the chance that it won't perform to specifications? As a project manager, how can I identify only the important risks to save time and money? As a program manager, how do I prioritize funds to identify, analyze, and control risk? How do I know when to stop? What's the residual risk? What's the best risk reduction strategy? Which idea, design, or concept has the best chance of success? What's the best purchase option?

PRA is a method for addressing the safety and reliability aspects of these questions—and many others like them—for high-consequence systems, projects, and programs that range from small and simple to large and complicated. Each question itself implies that a decision is to be made. PRA, then, is conducted in anticipation of decisionmaking. When we speak of risk management, we're talking about a process that uses risk assessment to make decisions about controlling risk. PRA is both a prelude to and a part of risk management and the associated decisionmaking because the methodology also yields much of the information needed to control risk.

3.1 Risk Management and PRA

We can think of the art of project management as managing uncertainties (read "risk") inherent in the conduct of a project. As the project begins and moves forward, a manager is, in effect, attempting to predict future outcomes when he makes decisions about how to achieve the mission goals of performance, cost, and safety. For example, when developing and designing a new Mars spacecraft (and the associated smaller spacecraft called "landers"), the ability of the spacecraft to reach the planet, deploy, land, and perform on the surface is uncertain.

The cost depends not only on the performance goals but also on the probability of mission success (i.e., the reliability) that is sought. Because we can't accurately predict future reliability, projects involve uncertainty. The project decisionmakers (DMs) must make decisions despite this uncertainty. Risk arises from uncertainty, and safety risk is associated with the uncertainty in being safe. Although high-consequence projects always involve risk, a DM typically has no tool or method for quantitatively estimating the safety risk. As a result, the DM is unable to balance safety and reliability against other important but uncertain attributes. PRA solves this problem.

3.2 Characterizing PRA

The primary purpose of PRA is to help DMs make good decisions about safety in a project or a system. Figure 3-1 is a generalized illustration of the performance of a system or project. In the figure, the line to the left of the word *perturbation* is a path of normal operation or normal progress of a project. After the perturbation (or disturbance), multiple paths can form, depending on the perturbation's characteristics and the system or project's internal controls. For example, the path of *continued normal operation* reflects system controls and feedbacks that reestablish stable operation.

The path labeled *controlled by shutdown* indicates that the perturbation caused a protection feature to shut down operation. The path labeled *severe consequences* indicates a large perturbation beyond the capability of all controls to deal with the problem. Many paths are possible; these are only three examples. And each path has a consequence, which in PRA is often termed the "end state."

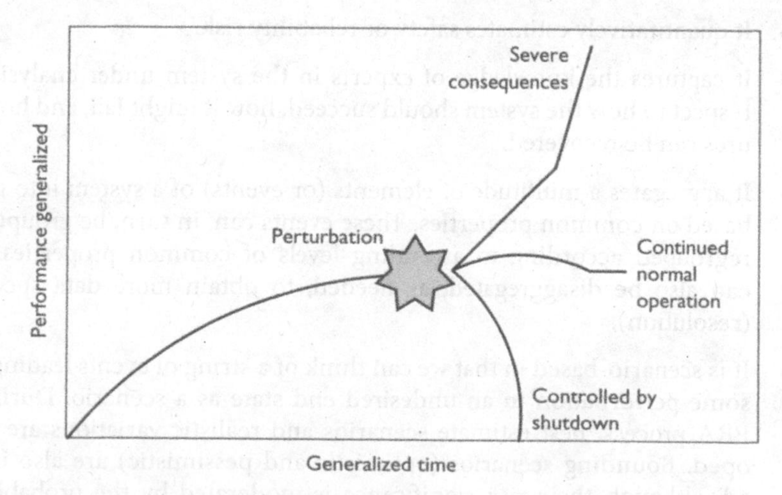

Figure 3-1. ***Concept of a Response to a Perturbation***

As we try to anticipate how a system might respond to a perturbation, a probabilistic perspective is essential. Each path has a probability of occurrence, and the probabilities themselves are often uncertain because of aleatory or epistemic reasons, or both. The severity of an end state (e.g., the number of injuries or deaths, the amount of property damage, or the cost of an accident) is often uncertain as well. By adding more detailed paths, we can conceptually reduce the uncertainty of the consequence of a specific path.[1] For example, we might imagine filling in the paths surrounding the severe consequences area in Figure 3-1 by adding paths that lead to one injury, two injuries, one death, multiple deaths, and so on.

We can view the conduct of PRA on a system as a study of the responses of that system to perturbations. Perturbations can come from any source: external environmental factors, interactions with people external to the system, or problems generated within the system (e.g., hardware, software, and human interaction problems). Simply put, a PRA simulates what happens when something goes wrong in a system's operation. The technique allows us to analyze the spectrum of end states that a system can cause to itself and its environment. It also permits us to estimate the probability of each state.

PRA combines probability theory and reliability engineering with other traditional engineering and scientific disciplines to develop structured, quantitative information suitable for aiding decisionmaking in the face of uncertainties. The result of a risk assessment is a risk model of the perturbations, end states, scenarios, and associated probabilities, with uncertainties, that can be used to estimate the safety risk and how to reduce it (if needed). To stay current with the project, a risk model remains fluid during a project. PRA uses techniques such as Bayesian data analysis, influence diagrams, event trees, fault trees, and Monte Carlo simulation. PRA's usefulness for making good decisions about safety derives from the following basic characteristics:

- It quantitatively estimates safety or reliability risk.
- It captures the knowledge of experts in the system under analysis with respect to how the system should succeed, how it might fail, and how failures can be recovered.
- It aggregates a multitude of elements (or events) of a system into groups based on common properties. These events can, in turn, be grouped and regrouped according to ascending levels of common properties. They can also be disaggregated, as needed, to obtain more data specificity (resolution).
- It is scenario-based in that we can think of a string of events leading from some perturbation to an undesired end state as a scenario. During the PRA process, best estimate scenarios and realistic variations are developed. Bounding scenarios (optimistic and pessimistic) are also included, although their risk significance is moderated by the probability of occurrence.

- It is structured, logical, and organized, allowing us to readily prioritize, for example, scenarios, elements, and component failures. The process elucidates the underlying reasons for the resulting order of priority. Because the path and causes of failure are developed, we gain insights into viable risk reduction strategies.
- It allows us to identify and quantify uncertainties associated with modeling of the physical, biological, and chemical aspects of events, the parameters of the models, and the probabilities or frequencies of events. Uncertainties can be propagated throughout the risk model so that the result includes all significant contributing uncertainties.

Any method that has these characteristics can be said to be a PRA. In the next section, I summarize some of the more popular methods.

3.3 Structure of a Traditional PRA[2]

A necessary prerequisite for a comprehensive and accurate PRA is a deep and broad knowledge of the system or project to be analyzed. PRA is typically performed by a team composed of individuals who have the PRA skills and those who are knowledgeable about the system. These prerequisites are necessary because PRA models how a system responds to perturbations, not all of which have been anticipated by the design engineers and operations personnel. PRA often requires engineering calculations (e.g., heat transfer, structural analysis, and electrical and control response simulation) to fully develop scenarios. The process of learning the system well enough to develop scenarios is called "system familiarization." During this process, emphasis is placed on establishing what PRA practitioners call "success criteria." The success criterion for a system or subsystem is the minimum complement of equipment functionality that still constitutes successful, albeit possibly degraded, performance. For example, spacecraft typically carry excess electric power production capability in the form of solar panels and battery cells, and scientific instruments can experience some sensor failure and still adequately provide data.

3.3.1 Scenarios

The basis of a risk assessment is the development of scenarios. We think of scenarios as strings of events that lead to an undesired consequence or end state. Each scenario begins with a trigger event, often called an "initiating event," and terminates with an "end state." Scenarios should include human actions and software errors, as well as hardware malfunctions; relevant environmental interactions; and relevant physical, biological, and chemical process phenomena. An initiating event is any perturbation, abnormality, malfunction, or failure (whether it is human, hardware, software, or process) that causes a deviation from desired operation. Examples of end states are mission loss, injury, system downtime, vehicle loss, nuclear reactor core damage, and toxic gas release. An end state can be a single undesired event, such as core melt or explosion, or an uncertain quantity, such as release of radioactive material, number of fatalities from an aircraft crash, or number of injuries from a flammable gas tank farm accident. End states are defined by the DM because the probability of them is used to make decisions.

In between trigger events and end states are "pivotal events" (sometimes called "intermediate events"), which determine whether and how an initiating event propagates to an end state. Figure 3-2 conceptually represents a scenario. As the figure shows, each scenario is defined by one initiating event, one or more pivotal events, and one end state.

Pivotal events can be aggravative, mitigative, protective, preventive, or benign. An aggravative event increases the severity of the end state, the frequency of reaching undesired end states, or both. A mitigative event reduces the severity of the end states. A protective event reduces the likelihood of the trigger event producing an undesired end state. A preventive event stops the end state from occurring. And a benign event has little or no effect on the course of the scenario, although the design engineers or PRA analysts may have perceived it as beneficial.

Scenarios can be developed and documented by a variety of diagrammatic forms, such as an event tree, a fault tree, a digraph, or an influence diagram. In fact, several forms can be used in the same risk assessment. PRA practitioners use these diagrams because (1) they lend themselves to quantification of scenario frequencies with uncertainties and (2) computerized construction and computational support is readily available for them. The diagram or other method is selected at the discretion of the analyst. When selecting diagrams, the analyst is influenced by the ease of model development and the display of results. Part of the "art" or creativity of performing a risk assessment is selecting the diagram-

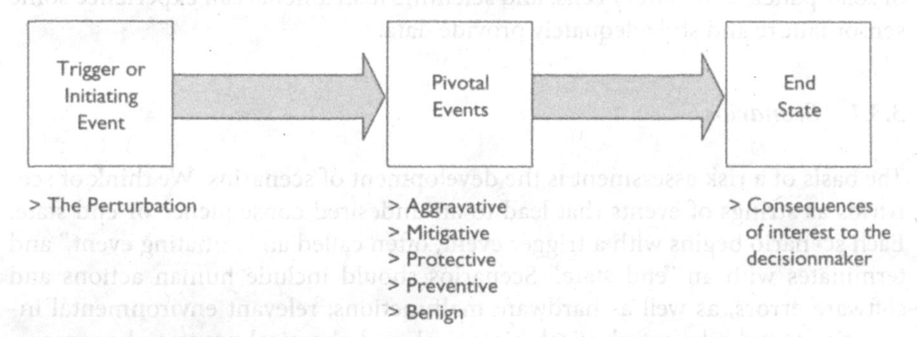

Figure 3-2. *Concept of a Scenario*

matic forms. The set of diagrams depends on the objectives and scope of the analysis as well as the audience for the results.

PRAs can also be performed with no diagrams at all. Some PRAs are solutions of first principle equations (e.g., equations of motion, heat transfer, fluid flow, and structural mechanics) that have identified the uncertainties and produced a probabilistic solution of results (see, for example, Lockheed Martin 1997; Frank et al. 2005). Scenarios may be characterized by complicated interrelationships of chemical, physical, and/or biological processes (e.g., entry body physics, hazardous material release physics, nuclear reactor core damage, and effects of toxic and radioactive materials). Their risk may be modeled using simulation, state vector methods, computation fluid dynamics, probabilistic explosion dynamics, structural mechanics, or other computational techniques.

3.3.2 *Defining the Set of Initiating Events*

Analysts start the risk assessment process by constructing a top-down overarching model of potential initiating events. Essentially any logical and hierarchical method that ensures completeness and allows an analyst to clearly think through and display his thoughts is acceptable. In the next paragraph, I briefly describe the most commonly used method, called a master logic diagram (MLD). Other methods use brainstorming, checklists, guidewords, failure mode and effects analysis, hazard and operability studies, and influence diagrams.

An MLD is a hierarchical method for developing a set of initiating events. It depicts ways in which system perturbations can occur. The boxes in an MLD are called "events," and the shape below each box is called a "gate."[3] Completeness in attempting to predict all such perturbations in every detail is quite impractical. A team of analysts, though, can usually capture all but the most indiscernible events by a systematic (e.g., functional) categorization of perturbations. An MLD starts with a top event that is an end state of interest to a DM who wants to improve the system. Referring to our tank farm example in Chapter 2, a DM may want to investigate future occurrences of loss of life, loss of property, fire, explosion, and/or tank burst, all of which are legitimate end states for investigation. Risk would be defined for such cases as the probability of explosion, loss of life, injury, or tank burst. An interpretation that explosions are "hazards" does not interfere with using an explosion as an end state, an initiating event, or a pivotal event in a PRA. Because a PRA is structured to meet the needs of a DM, its structure must only be self-consistent toward that goal. In other words, there are no a priori constraints on the types of events used for end states, initiating events, and pivotal events.

After the DM establishes the top event, the PRA analyst asks herself, "How can the top event occur? Events that are *necessary but not sufficient* to cause the top event are enumerated in ever more detail as lower levels of the hierarchy are built. In other words, the MLD depicts only the *necessary* initiating events but not the *sufficient conditions* (i.e., pivotal events) that cause the end state of a scenario. Because the MLD is not intended to be a complete depiction of all combinations of events or scenarios that cause all end states, the diagrams usually contain a predominance of OR gates. An MLD must be constructed for each end state of interest.

Typically, as illustrated in Figure 3-3, the top levels are functional failures (e.g., failure to cool, failure to relieve pressure, failure to isolate, or fire). The lower levels are subsystem and component failures that contribute to the functional failures. The diagram continues toward more detailed events as long as each identified initiating event category has more than one system response. For example, suppose the DM decides to investigate the risk associated with single or multiple tank bursts in a flammable liquid gas tank farm. One downward flow of logic in Figure 3-3 would be associated with fire at Level 1 (functional events), and others might be overfilling or loss of spray. Looking into the fire event further, Level 2 (system level) would consist of, for example, inadvertent relief or drain opening. Level 3 would, for example, include events such as drain valve leak, pipe leak, relief valve leak, or drain valve failing to open. Level 4 would include such events as valve A remains open, valve C remains open, and so forth. Ultimately, a level of detail in the downward logic path will be reached such that enumerated events have the same system response. Initiating events are those for which the same pivotal events and end states would be appropriate. For example, it may not matter whether a leak comes from valves A and C or valves A and B or a pipe leak (refer to Figure 2-2). In fact, it may not matter at all how a fire started because all subsequent events would be quite similar, irrespective of the fire's cause. The Level 1 event fire, therefore, would be a legitimate initiating event. Its causes would be delineated and analyzed as part of a fault tree, which I describe later in this chapter.

Following the development of an MLD or another equivalently comprehensive method to generate an exhaustive set of initiating event categories, PRA practitioners typically select another diagram to characterize how the system responds. Common diagrams for this are event sequence diagrams or event trees—both elaborate on how the system responds to an initiating event by delineating the scenarios. These diagrams show how pivotal events combine with initiating event to lead to the end state. Using fault trees to disaggregate initiating and pivotal events into their causal events is often convenient, as I discuss in the next sections.

3.3.3 *Event Sequence Diagrams and Event Trees*

An overview of a system's response to an event is typically depicted in a diagram called an event sequence diagram (ESD). An ESD represents scenarios in terms of initiating events, pivotal events, and end states. Because the future is uncertain, the analyst does not know which of the alternative scenarios might occur. This is analogous to a decision analysis (refer to Figure 1-1), in which the DM does not know which alternative will yield the most favorable outcome. In an ESD, uncertainty is explicitly shown by attempting to identify all reasonable alternative scenarios and their end states. In a decision, all reasonable alterna-

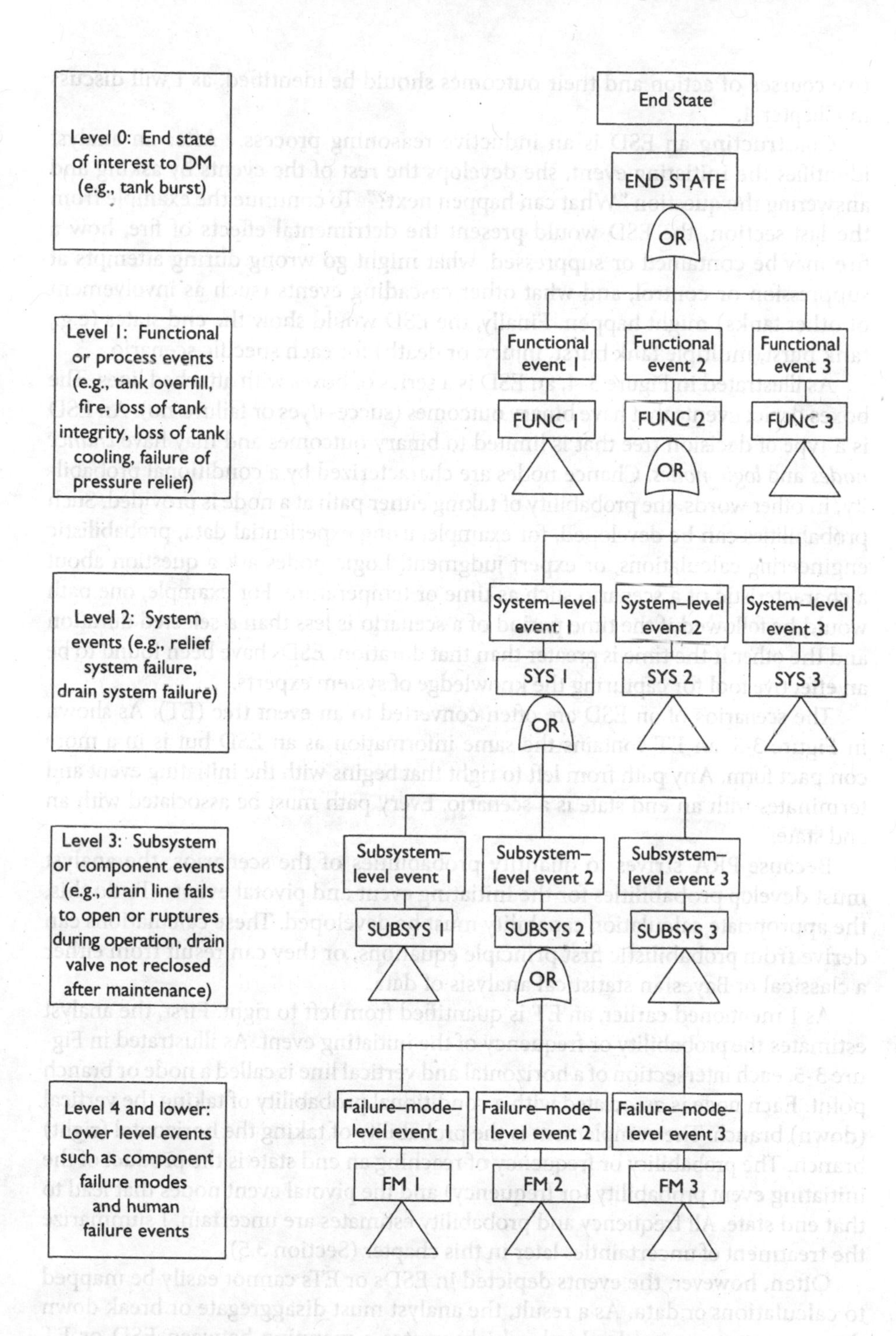

Figure 3-3. ***Concept of a Master Logic Diagram***

tive courses of action and their outcomes should be identified, as I will discuss in Chapter 4.

Constructing an ESD is an inductive reasoning process. After an analyst identifies the initiating event, she develops the rest of the events by asking and answering the question "What can happen next?" To continue the example from the last section, the ESD would present the detrimental effects of fire, how a fire may be contained or suppressed, what might go wrong during attempts at suppression or control, and what other cascading events (such as involvement of other tanks) might happen. Finally, the ESD would show the end states (e.g., tank burst, multiple tank burst, injury, or death) for each specific scenario.

As illustrated in Figure 3-4, an ESD is a series of boxes with attached lines. The boxes depict events that have binary outcomes (success/yes or failure/no). An ESD is a type of decision tree that is limited to binary outcomes and may have *chance nodes* and *logic nodes*. Chance nodes are characterized by a conditional probability; in other words, the probability of taking either path at a node is provided. Such probabilities can be developed, for example, using experiential data, probabilistic engineering calculations, or expert judgment. Logic nodes ask a question about a characteristic of a scenario such as time or temperature. For example, one path would be followed if the time period of a scenario is less than a selected duration and the other if the time is greater than that duration. ESDs have been found to be an effective tool for capturing the knowledge of system experts.

The scenarios of an ESD are often converted to an event tree (ET). As shown in Figure 3-5, an ET contains the same information as an ESD but is in a more compact form. Any path from left to right that begins with the initiating event and terminates with an end state is a scenario. Every path must be associated with an end state.

Because PRA strives to quantify probabilities of the scenarios, the analyst must develop probabilities for the initiating event and pivotal events. To do this, the appropriate calculation capability must be developed. These calculations can derive from probabilistic first principle equations, or they can result from either a classical or Bayesian statistical analysis of data.

As I mentioned earlier, an ET is quantified from left to right. First, the analyst estimates the probability or frequency of the initiating event. As illustrated in Figure 3-5, each intersection of a horizontal and vertical line is called a node or branch point. Each node is associated with a conditional probability of taking the vertical (down) branch. The complement is the probability of taking the horizontal (right) branch. The probability or frequency of reaching an end state is the product of the initiating event probability (or frequency) and the pivotal event nodes that lead to that end state. All frequency and probability estimates are uncertain. I summarize the treatment of uncertainties later in this chapter (Section 3.5).

Often, however, the events depicted in ESDs or ETs cannot easily be mapped to calculations or data. As a result, the analyst must disaggregate or break down these events to a simpler level, which creates a mapping between ESD or ET events and the available data or calculations. In risk assessment, the most popular form of this mapping is a fault tree (FT).

3.3.4 *Fault Trees*

As illustrated in Figure 3-6, FTs are often useful in developing higher resolution (or more detailed) events within an ESD or an ET. Construction of an FT is a deductive reasoning process that answers the question, "What are all combinations of events that can cause the top event to occur?" When top events are used in risk assessment, they can be either a pivotal or an initiating event. This top-down analytical development defines the combinations of causes of the events in a scenario in a way that allows the probability of the events to be estimated.

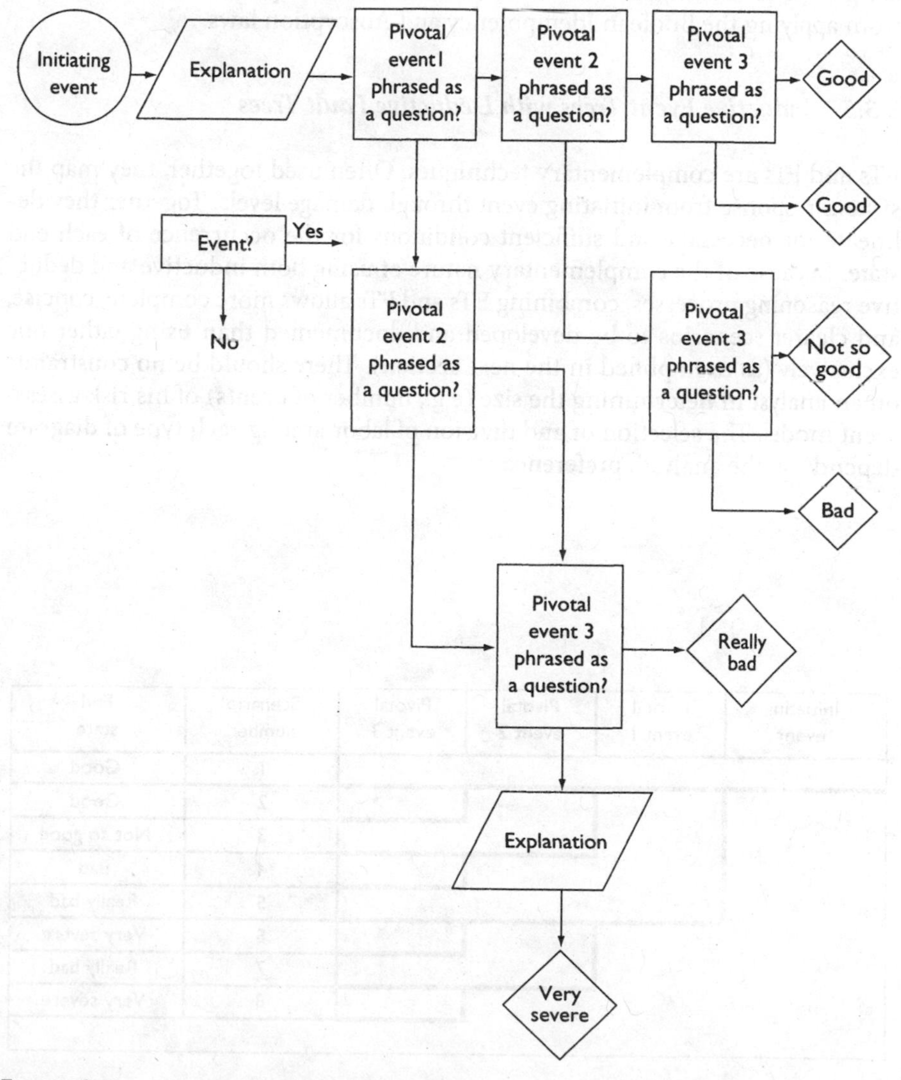

Figure 3-4. *Event Sequence Diagram*

FTs use logic or Boolean gates. Figure 3-6 shows two types of gates: AND and OR gates. An AND gate passes an output up the tree if all events immediately attached to it occur. An OR gate passes an output up the tree if one or more events immediately attached to it take place. As the name implies, FT events are usually failures or faults. An AND gate often implies components or system features that back each other up, so that if one fails the other continues to adequately perform the function. Knowledge of system success criteria, as I discussed earlier, is essential for determining the type of gate when using an FT to analyze a system.

FTs can be Boolean reduced to "minterm" form, which expresses the top event in terms of the union or "sum" of minimal cutsets. Minimal cutsets, which are groups of events that must all occur to cause the top event in the FT, result from applying the Boolean Idempotency and Absorption laws.

3.3.5 *Inductive Event Trees with Deductive Fault Trees*

ETs and FTs are complementary techniques. Often used together, they map the system response from initiating event through damage levels. Together they delineate the necessary and sufficient conditions for the occurrence of each end state. Because of the complementary nature of using both inductive and deductive reasoning processes, combining ETs and FTs allows more complete, concise, and clearer scenarios to be developed and documented than using either one exclusively (as exemplified in the next section). There should be no constraints on an analyst in determining the size (e.g., number of events) of his risk assessment model. The selection of and division of labor among each type of diagram depends on the analyst's preference.

Initiating event	Pivotal event 1	Pivotal event 2	Pivotal event 3	Scenario number	End state
				1	Good
				2	Good
				3	Not so good
				4	Bad
				5	Really bad
				6	Very severe
				7	Really bad
				8	Very severe

Figure 3-5. *Event Tree*

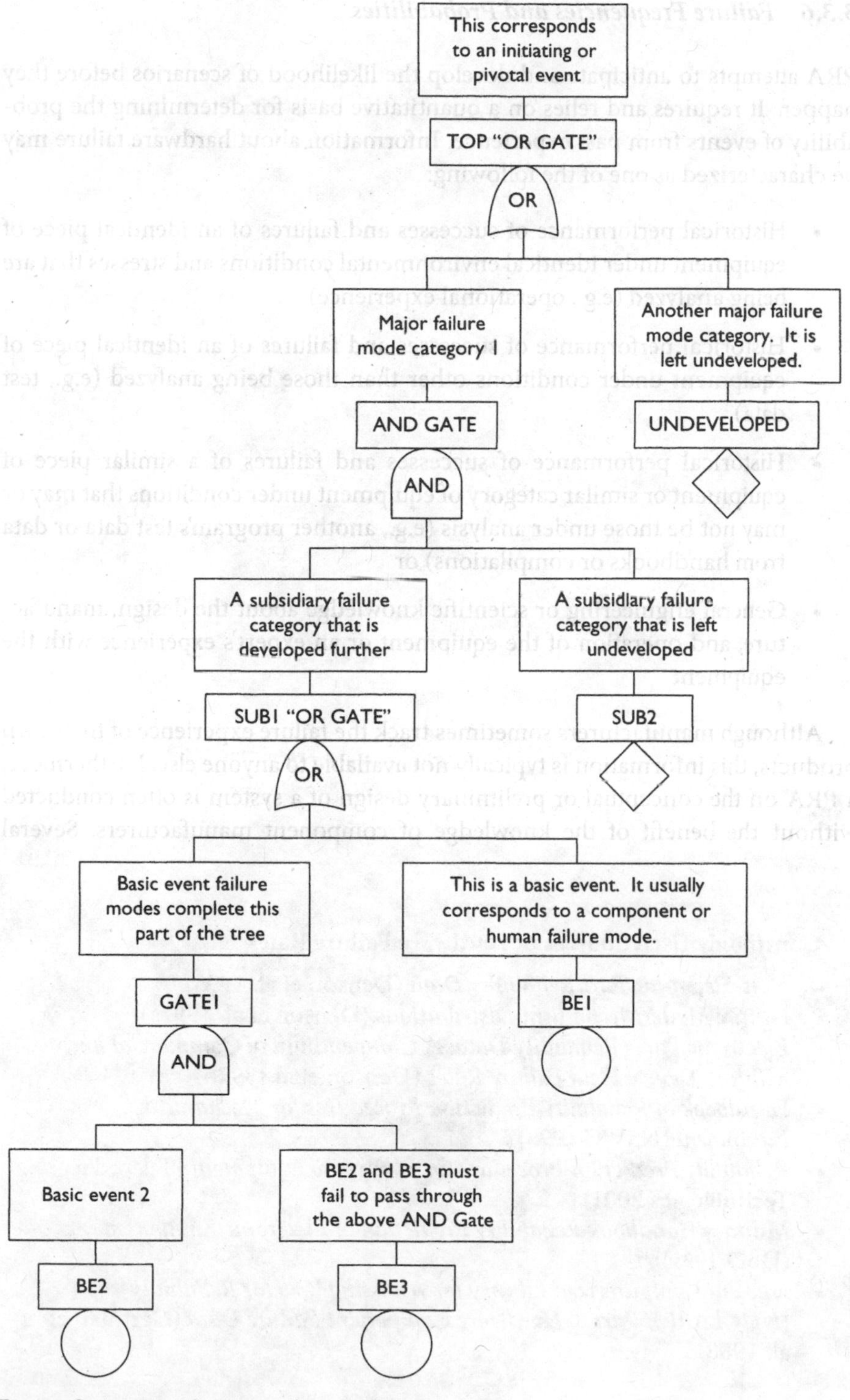

Figure 3-6. *Fault Tree*

3.3.6 *Failure Frequencies and Probabilities*

PRA attempts to anticipate and develop the likelihood of scenarios before they happen. It requires and relies on a quantitative basis for determining the probability of events from past experience. Information about hardware failure may be characterized as one of the following:

- Historical performance of successes and failures of an identical piece of equipment under identical environmental conditions and stresses that are being analyzed (e.g., operational experience)
- Historical performance of successes and failures of an identical piece of equipment under conditions other than those being analyzed (e.g., test data)
- Historical performance of successes and failures of a similar piece of equipment or similar category of equipment under conditions that may or may not be those under analysis (e.g., another program's test data or data from handbooks or compilations) or
- General engineering or scientific knowledge about the design, manufacture, and operation of the equipment or an expert's experience with the equipment

Although manufacturers sometimes track the failure experience of their own products, this information is typically not available to anyone else. Furthermore, a PRA on the conceptual or preliminary design of a system is often conducted without the benefit of the knowledge of component manufacturers. Several

Commonly Used Sources of Hardware Failure Rates

- *Non-Electronic Part Reliability Data* (Denson et al. 1995)
- *Failure Mode/Mechanism Distributions* (Denson et al. 1997a)
- *Electronic Parts Reliability Data: A Compendium of Commercial and Military Device Field Failure Rates* (Denson et al.1997b)
- *Handbook of Reliability Prediction Procedures for Mechanical Equipment* (NSWC 1998)
- *Reliability Prediction Procedure for Electronic Equipment* (Telcordia Technologies 2001)
- *Military Handbook Reliability Prediction of Electronic Equipment* (DoD 1991)
- *Nuclear Computerized Library for Assessing Reactor Reliability (NUCLARR), Part 3: Hardware Component Failure Data* (Gertman et al. 1988)

commercial hardware failure rate sources are publicly available and several more can be found in reliability engineering books (see shaded box on page 40).

3.3.7 *Illustrative Tank Farm PRA Example*

In this section, I use a simple example to illustrate the PRA structure and quantification introduced in the previous sections. In our example, a company was considering constructing a flammable liquid propane gas (LPG) tank farm (see Figure 3-7 for the proposed tank farm layout). Wanting to know what risks the tank farm would pose, a public interest group hired a consulting firm to do a PRA.

Discussions between the public interest group and the consulting firm led to the decision to perform a conservative analysis of risk that focused on the tank farm equipment itself. For this study risk was defined as the probability of at least one tank burst over a 10-year period. The consultants chose this definition because it would yield an estimate of the probability that local and regional

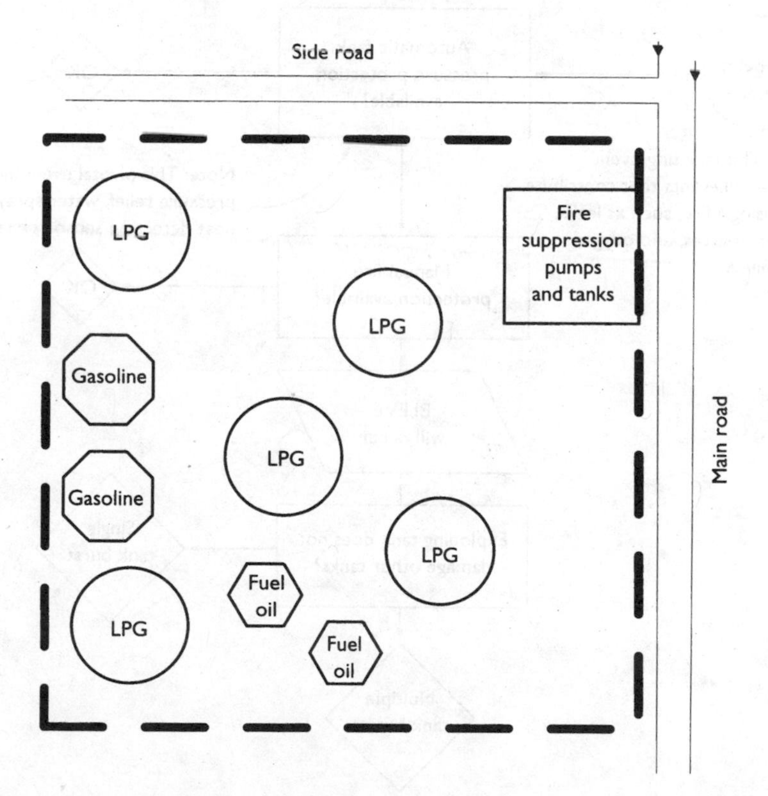

Figure 3-7. ***Proposed Tank Farm***

Notes: LPG = liquid propane gas. This depiction is illustrative only. It does not represent the Feyzin event.

authorities would have to deal with a disaster. After hearing about the study's focus, the gas storage company decided to join in the study—not only for public relations reasons but also because such an estimate of risk would be a useful measure of the risk of its investment.

In the proposed design, each tank has a relief valve and a set of drain valves (refer to Figure 2-2), along with a leak detection system that can detect leaks through the relief valve and drain valves. The tanks are also equipped with water spray that is initiated by temperature sensors on the tank surface. Analysts assumed that maintenance personnel would drain the tanks once per month.

The consulting firm developed an MLD (as summarized in Section 3.3.2) for the tank burst end state. One of the initiating events was "fire below tank." The ESD in Figure 3-8 presents the scenarios the firm developed, which would emanate from the initiating event. The diagram shows that if either pressure suppression or fire protection were available, the fire would do no significant damage, as indicated by the "OK" end state. If neither were available, the down branch in the

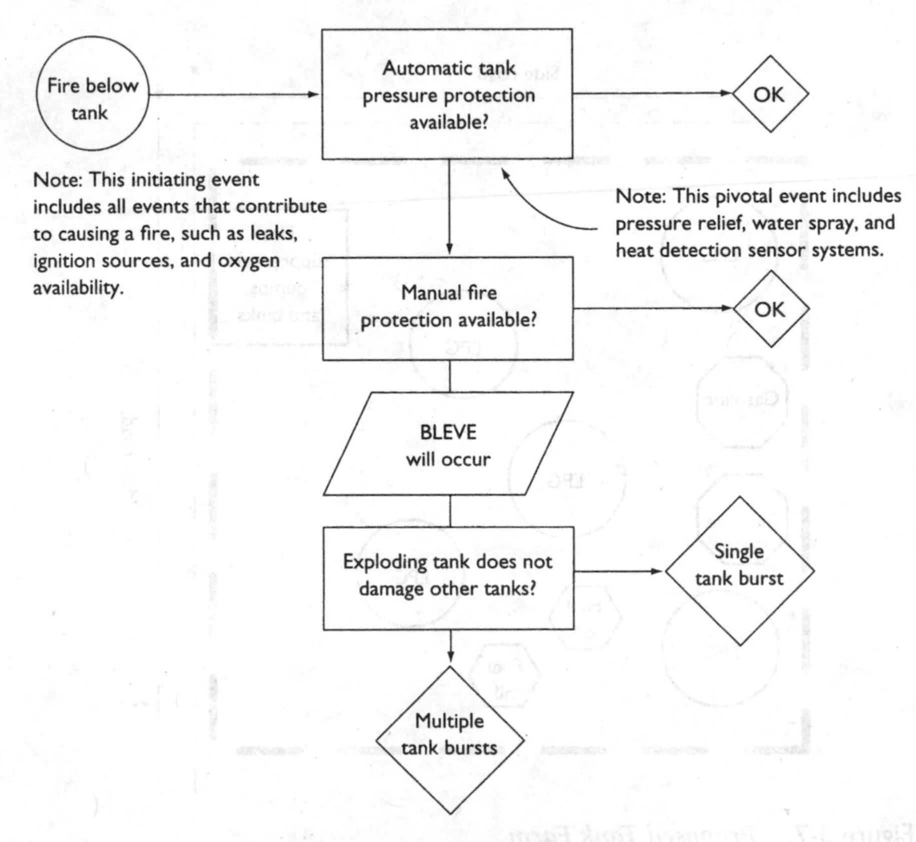

Figure 3-8. ***Event Sequence Diagram for a Tank Fire***
Note: BLEVE = boiling liquid expanding vapor explosion.

figure indicates that a BLEVE would occur, possibly damaging multiple tanks. Depending on the severity of the BLEVE and the proximity of the other tanks, either the multiple tank bursts or the single tank burst end state would apply.

Figure 3-9 shows the equivalent ET for the scenarios in Figure 3-8. The OK end state is a euphemism. Even though scenarios 1 and 2 would cause some damage, PRA practitioners commonly assign OK to a scenario end state that is not of interest to the study.

In the figure, the initiating event probability, F, is the probability of a fire below a tank over 10 years of tank farm operation. By convention, the pivotal event lower branch denotes failure or unavailability. The conditional probability of this failure or the unavailability of the pivotal events are given in Figure 3-9 as A (automatic tank pressure protection fails), M (manual fire protection fails), and E (the exploding tank damages other tanks). Success or availability is denoted by the complement. Consider, for example, the conditional probability of automatic tank pressure protection availability, 1-A. Within the context of an ET, the fraction of occurrences that follow the lower branch is A. The probability of a scenario is the product of the probabilities of the events in a path. The figure shows four scenarios, and the probability of each is given in the column labeled Probability. For example, Scenario 4 comprises F, A, M, and E, so the probability of this scenario is FAME. Because the ET merely partitions the initiating event probability into alternative scenario probabilities, the probabilistic sum of all scenarios in an ET must equal the initiating event probability, which is F in this example.

To develop the probabilities F, A, M, and E and to understand the causes of these events, the consultants used FTs. Figure 3-10 illustrates an FT that disaggregates the event "fire below LPG tank" into its causal events. To cause a fire below the tank in the example, propane would have to leak, the mixture would have to reach a flammable mixture, and the mixture would have to be ignited. All these events took place during the Feyzin accident I described in Chapter 2.

The FT breaks apart the event "fuel below tank" into individual actions, such as "crew forgets to close valve," and individual hardware malfunctions, such as "valve A fails to close."[4] Because these consultants were skillful PRA analysts,

Fire below LPG tank	Automatic tank pressure protection available?	Manual fire protection available?	Exploding tank does not damage other tanks?	End state	Probability
F	A	M	E		
				OK	F(1-A)
				OK	FA(1-M)
				Single tank burst	FAM(1-E)
				Multiple tank burst	FAME

Figure 3-9. ***Event Tree for a Tank Fire***

Notes: LPG = liquid propane gas. Refer to text for explanations of F, A, M, and E.

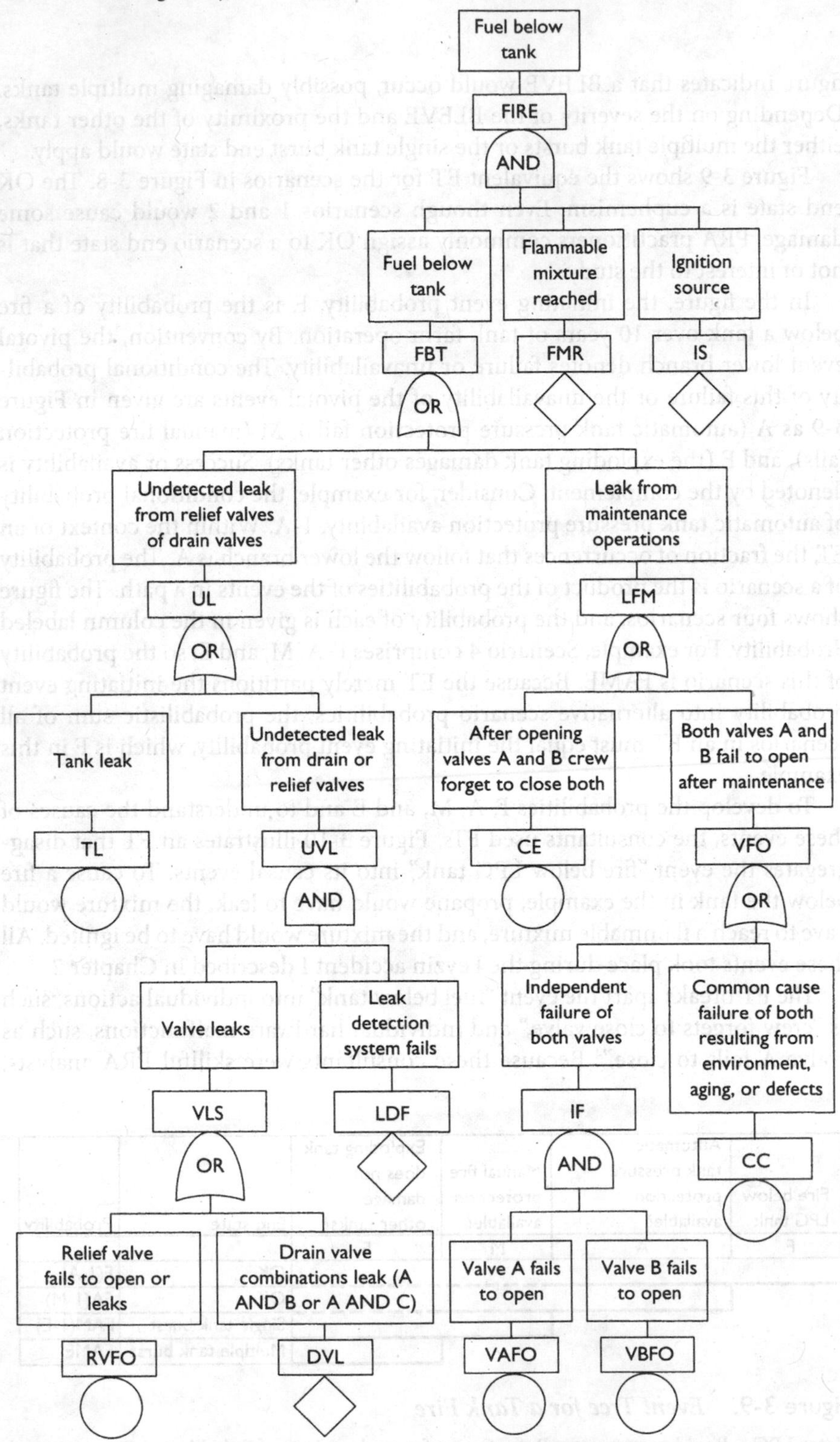

Figure 3-10. ***Initiating Event: Fire Below Liquid Propane Gas Tank***

they delineated the causes of the top event so that the basic events corresponded to events for which probabilities (e.g., failure rate, unavailability, and unreliability) were available or easily developed.

The consultants also used FTs to further analyze the pivotal event "automatic tank pressure protection fails," as shown in Figure 3-11.

In Figure 3-11, an "undeveloped event" is depicted by a diamond at the bottom of an event. It denotes a situation in which the event is usually disaggregated into smaller parts. In this case, however, the consultants elected not to break down the event further because event probability information was available for the aggregate. A "basic event" is illustrated in Figures 3-10 and 3-12 by a circle at the bottom of an event. It denotes a situation in which, by convention, the event is not broken down further because event probability information is typically available for such an event.

To obtain the desired risk (i.e., the probability of one or more tank bursts over 10 years), the consultants quantified the model, as I describe next. The consultants populated each of the basic events and undeveloped events in the FTs with probabilities in the manner I described in Sections 3.3.6, 3.4.1, and 3.4.2. So as to expeditiously show an overview of the PRA structure and quantification, I do not specifically describe how the consultants developed these probabilities. Although each of the event probabilities is uncertain and represented by a probability distribution, this simplified example shows a single number for each estimate.[5] I discuss quantification with probability distributions in following sections.

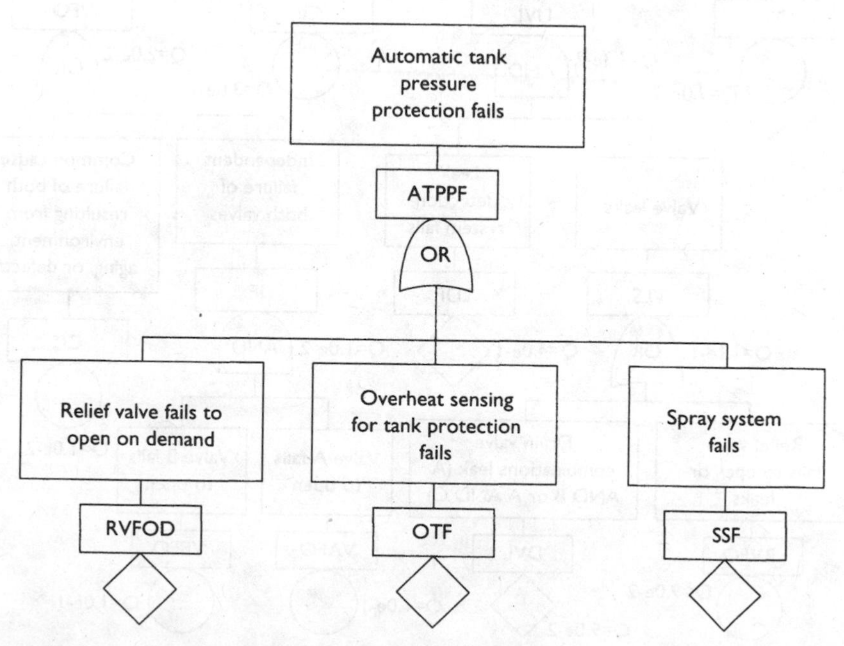

Figure 3-11. ***Fault Tree for Pivotal Event: Automatic Tank Pressure Protection Fails***

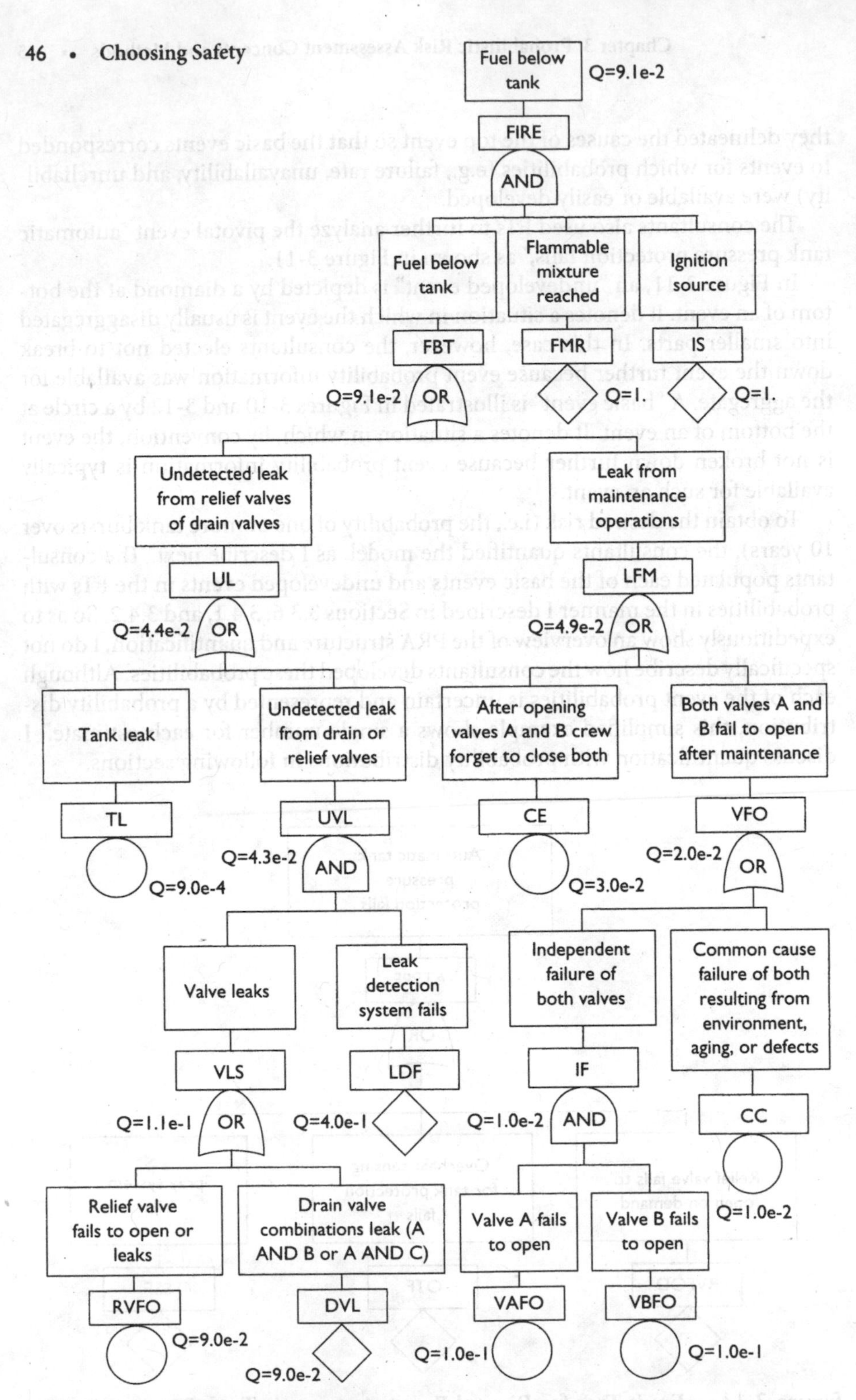

Figure 3-12. ***Quantification of Initiating Event Fault Tree for 10-Year Probability of Fire Below Tank***

Note: Q = probability or unavailability.

The consultants *developed* the PRA model from the top down, but they *quantified* it from the bottom up. As Figure 3-12 shows, probabilities are associated with basic events and undeveloped events. Here, the consultants combined the event probabilities using probabilistic addition under OR gates and probabilistic multiplication under AND gates.

The quantification reflected the consultants' finding of a high likelihood of an ignition source being present somewhere within a few hundred yards of the tank farm and a flammable mixture eventually occurring. Prudently, the consultants assigned a probability of 1 to each of these events, which brings the 10-year probability of having a fire below a tank to approximately 1/11. Five LPG tanks were to make up the farm, and each was taken to have the same probability of fire. Similarly, the consultants populated the pivotal event FT for "automatic tank pressure protection fails" with probabilities and quantified it as shown in Figure 3-13.

Local fire brigades typically train in response to a variety of postulated fire scenarios. These fire drills are usually timed so that data are available about response times before a tank explodes. Using such data, the consultants were able to develop computer simulations of vessel fires that included estimates of the times to vessel rupture under a variety of response times. They employed Monte Carlo simulation techniques coupled with the appropriate vessel thermal and pressure equations to quantify the pivotal event probability for "manual fire protection available."

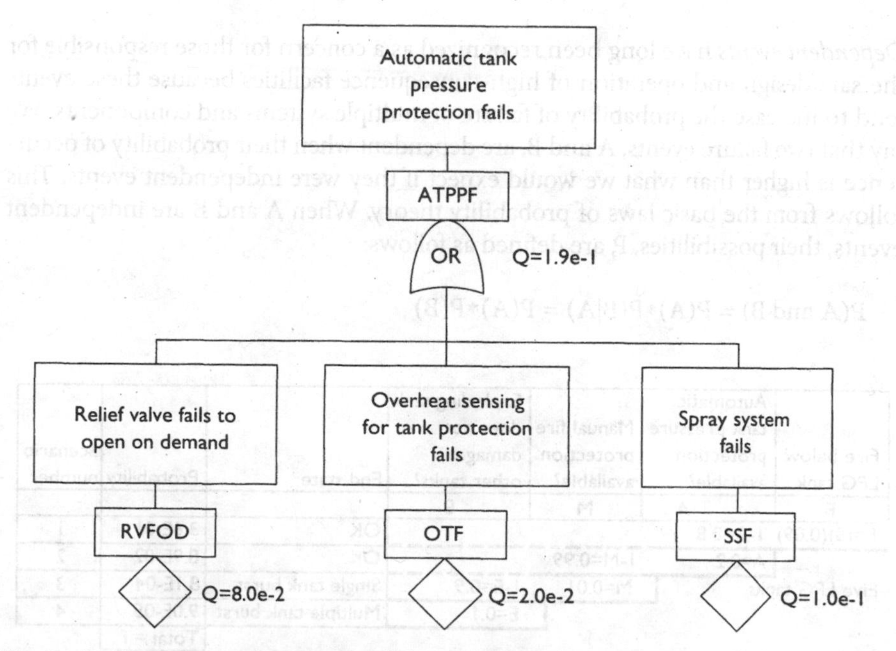

Figure 3-13. ***Pivotal Event Fault Tree Quantification***

Note: Q = probability or unavailability.

Finally, they used coupled explosion dynamics and structural analysis software to estimate damage propagation from an exploding body to adjacent structures. In this manner, the pivotal event probability of "Exploding tank does not damage other tanks?" is quantified.

Figure 3-14 shows the completed estimate of the probability of the tank farm fire scenarios.

In our example, the consultants arrived at an estimate of the total probability of one or more tanks bursting over a 10-year period of approximately 1/1,111. In other words, the odds against a tank burst in 10 years were 1,110 to 1. When we consider the many tank farms around the world, these are not particularly reassuring odds for a single tank farm.

In looking at the example analysis, we can see that the best way to improve safety at the site would be to prevent single tank bursts. The example ET shows that the best way to do this would be to increase the reliability of automatic tank pressure protection and reduce the probability of leaks. As we dig deeper, the FTs illustrate that the largest contributor to failure of the automatic tank pressure protection would be the single-string fire-spray system. Finally, the FTs show that the largest contributors to leaks are maintenance actions and failure to detect a leak.

3.4 Dependent Events

Dependent events have long been recognized as a concern for those responsible for the safe design and operation of high-consequence facilities because these events tend to increase the probability of failure of multiple systems and components. We say that two failure events, A and B, are dependent when their probability of occurrence is higher than what we would expect if they were independent events. This follows from the basic laws of probability theory. When A and B are independent events, their possibilities, P, are defined as follows:

$$P(A \text{ and } B) = P(A)*P(B|A) = P(A)*P(B)$$

Fire below LPG tank	Automatic tank pressure protection available?	Manual fire protection available?	Exploding tank does not damage other tanks?	End state	Probability	Scenario number
F	A	M	E			
F=(5)(0.09)	1-A=0.8			OK	3.6E-01	1
	A=0.2	1-M=0.99		OK	8.9E-02	2
Five LPG tanks		M=0.01	1-E=0.9	Single tank burst	8.1E-04	3
			E=0.1	Multiple tank burst	9.0E-05	4
					Total = F	

Figure 3-14. *Completed Event Tree Quantification*

When B depends on A, however,

$$P(A \text{ and } B) = P(A)*P(B|A) \neq P(A)*P(B)$$

When $P(B|A) >> P(B)$, safety concern may be heightened.

In PRA, we can divide dependent events into four dependence mechanisms: functional, environmental, spatial, and human. PRA practitioners devote a large fraction of the work required to develop and maintain a comprehensive PRA model to identifying and analyzing dependent events.

Functional dependence is present when one component or system relies on another to supply vital functions. Examples are the electric power supply to thermal control systems, thermal controls for computer systems, and computer controls of equipment and processes. Or, part of a system may be unavailable because of an outage (e.g., for maintenance or testing). Functional dependence is explicitly modeled in the ET and FT logic.

Environmental dependence is in play when system functionality relies on maintaining an environment within designed or qualified limits. Here, examples are material property and thermal stress changes associated with temperature changes, corrosion enhanced by stress, or motor degradation caused by high dust or humidity or both. We model these as modifications to system and component-failure probabilities. Other events, such as earthquakes, lightning strikes, high winds, fires, floods, and soil subsidence, can degrade multiple structures, systems, and components. We model these explicitly as initiating events in ETs.

Spatial dependence is at work when one structure, system, or component fails by virtue of close proximity to another. For example, an electrical wire within a fuel tank can ignite fuel vapors; an earthquake or tornado can cause walls to break apart, which in turn causes equipment supported by the wall and located next to the wall to fail; fire in an electrical cabinet or in lube oil can cause overhead cables or equipment to fail; failure of the pressurized compartment in an aircraft can cause floor movement and rupture of attached hydraulic lines; and unplanned operation of water sprinklers can cause the electrical equipment below to fail. We discover spatial dependences by explicitly looking for them in the layout of a facility and modeling these dependences in the ETs and FTs.

Human dependence is present when a structure, system, component, or function fails because humans intervene or failed to intervene. Here, examples include human errors in decisionmaking, design, manufacturing, assembly, diagnostics, inspections, procedure adherence, maintenance (or its absence), or inadequate equipment qualification. In a PRA, we treat these using the specialty of human reliability analysis within the ET and FT logic. We may also include these dependent events within the failure probability.

Common cause events can result from any of these examples. The term *common cause event* is widely employed to describe events in which the same cause degrades the function of two or more components, either at the same time or within a short time relative to the overall component mission time. When the extent of degradation of the components involved in the event progresses to the point of producing multiple failures in parallel redundant portions of a system, we call this event a *common cause failure* (CCF). Because of their high level of risk significance, CCFs are a special class of dependent failures that we emphasize in developing and quantifying PRA models.

Although relatively rare, CCFs have been instrumental in some major accidents:

- Hydrazine leaks leading to two auxiliary power unit (APU) explosions on the space shuttle (STS-9, 1983)
- Multiple aircraft engine failures (e.g., Fokker F27, 1988, 1997; Boeing 747, 1992)
- Three hydraulic system failures following engine #2 failure on DC-10, 1989
- Two auxiliary feedwater system valves found inadvertently closed during the 1979 Three Mile Island accident,
- One of two space shuttle main engine controllers on two separate engines failed when a wire shorted (STS-93, 1999)
- Failure of two O-rings causing hot gas blow-by in a solid rocket motor of space shuttle flight 51L (the *Challenger* accident, 1986)

Because CCFs are relatively uncommon, it is difficult to develop a statistically significant sample when we're monitoring only one system or facility, or even several systems. The development of CCF techniques and data, therefore, relies on a national data collection effort that monitors a large number of nuclear power systems. Typically, the fraction of component failures associated with common causes leading to multiple failures ranges between 1% and 10% (Fleming and Rao 1992; Marshall et al. 1998a,b; Mosleh et al. 1998; Rasmuson et al. 1998). This fraction depends on the component; level of redundancy (e.g., two, three, four); duty cycle; operating and environmental conditions; maintenance interventions; and testing protocol, among others. For example, equipment that is operated in cold standby mode (i.e., called to operate occasionally on demand) with a large amount of preventive maintenance intervention tends to have a higher fraction of CCFs than systems that continuously run.[6]

When we perform a PRA, it is not practical to explicitly identify all CCFs. Surveys of failure events in the nuclear industry have led to several parameter models. Of these, three are most commonly used: the Beta Factor method (Fleming 1975), the Multiple Greek Letter (MGL) method (Fleming and Kalinowski 1983), and the Alpha Factor method (Mosleh et al. 1988). These methods, which I describe in the following sections, do not require an explicit knowledge of the dependence failure mode. If we do know the specific failure mode, though, we model it explicitly instead of using these parameter methods.

We use the MGL method for common cause groups of up to four. A common cause group (CCG) is a set of redundant identical components that all perform the same system function.

The MGL parameters consist of a set of failure fractions used to quantify the conditional probabilities of all the possible ways a CCF of a component can be shared with other components in the same group, once a component has failed. A CCG is defined for each relevant failure mode. For a system of CCG size *m* (where *m* is four or less), let

Q_t = total single-component failure probability, which accounts for both independent and CCF modes affecting that component;

$\beta^{(m)}$ = beta, the fraction of the component failures involved in CCFs with other components, a function of the size of the CCG, *m*;

$\gamma^{(m)}$ = gamma, the fraction of the component CCFs that is shared with at least two other components, a function of the size of the CCG, *m*; and

$\delta^{(m)}$ = delta, the fraction of the component CCFs shared with at least two other components that are shared with three other components (i.e., all components in a group of four), a function of the size of the CCG, *m*.

When we apply the MGL model with more than four components in a CCG, we normally assume that any common cause event that would fail four components would fail all components in the group. The equation that expresses the probability of multiple-component failures resulting from a common cause, $Q_k^{(m)}$, is

$$Q_k^{(m)} = \frac{1}{\begin{pmatrix} m-1 \\ k-1 \end{pmatrix}} \prod_{i=1}^{k} \rho_i \; (1 - \rho_{k+l}) \, Q_t$$

$$\rho_1 = 1, \rho_2 = \beta, \; \rho_3 = \gamma, \rho_4 = \delta \ldots \rho_{m+1} = 0$$

where *m* is the number of components in the common cause group, and *k* is the number of specific components that fail such that $1 < k < m$. The binomial term

$$\begin{pmatrix} m-1 \\ k-1 \end{pmatrix} = \frac{(m-1)!}{(m-k)!(k-1)!}$$

represents the number of different ways that a specific component can fail. If $m = 2$, the model reduces to the Beta Factor method.

The Alpha Factor method was developed to address a limitation of the MGL method that Mosleh discovered arises while performing uncertainty analysis (Mosleh et al. 1988). This limitation is created by the mutual interdependence of the MGL parameter definitions, which tends to result in a slight understatement of the variance of the resulting uncertainty distributions. The Alpha Factor method is similar to the MGL method, except that instead of expressing fractions of failures in a component that result from a common cause, it uses fractions of failure that occur in a system.

Let $\alpha_k^{(m)}$ = a parameter of the kth order CCF where $k = 1 - 4$ for a CCG of size m. Then, the equation that expresses the probability of multiple-component failures resulting from a common cause, $Q_k^{(m)}$ is

$$Q_k^{(m)} = \frac{m}{\binom{m}{k}} \frac{\alpha_k^{(m)}}{\alpha_t} Q_t$$

where

$$\alpha_t^{(m)} = \sum_{k=1}^{m} k \alpha_k^{(m)}$$

and

$Q_k^{(m)}$ = probability of the kth order common cause failure for group m.

3.5 Uncertainties

In PRA, the fundamental viewpoint is probabilistic. The complicated nature of the potential scenarios demand that we account for both natural variability of physical processes (i.e., aleatory or stochastic uncertainty) and the uncertainties in knowledge of these processes (i.e., epistemic or state-of-knowledge uncertainty). Aleatory uncertainty refers to the inherent variation of a physical process over many similar trials or occurrences. Wind direction and fuel-explosion burn temperature are two very different example processes that exhibit variability. Weather is an obvious example of aleatory uncertainty because it changes constantly as atmospheric conditions change.

Consider the mind experiment in which a highly flammable and explosive fuel and oxidizer begin to mix, are ignited, and hypothetically undergo repeated explosions. We would expect the fireball's temperature to vary because of the stochastic nature of burn environments and processes. To explain further, fireball temperature is a function of many variables, such as fuel composition, ignition delay to allow mixing of fuel and oxidizer, rate of mixing, distance from ground, and turbulence effects. Each of these is subject to random variations, and they are, therefore, considered stochastic variables. These variations are sometimes called parameter uncertainties. Moreover, there is no exact mathematical fire burn and explosion model, although there are reasonable approximations. The uncertainty in a mathematical model is often called *model uncertainty.* Much of such a model's lack of accuracy derives from our incomplete understanding of the interactions of the stochastic variables.

Model uncertainty is an example of epistemic uncertainty (recall that epistemic uncertainty refers to our state of knowledge about a parameter or model). If we can improve our state of knowledge, arriving at more accurate information about a parameter, then the uncertainty is reducible. Epistemic uncertainty, therefore, is sometimes called reducible uncertainty. For example, we could paint an accurate picture of a particular tank farm component's failure rate if we could perform a sufficient number of operational trials (usually on the order of hundreds of thousands to millions of trials for typical equipment). Other uncertainties arise from, for example, inaccuracies in modeling and data, the applicability of data to the situation of interest, and our incomplete knowledge of the physical processes at work. Many studies have shown us that the uncertainties associated with using available experimental data, selecting calculation models, using simplifying assumptions, and varying quantities of variables used as input to the calculations are important sources of uncertainty within a PRA.

Any particular scenario may or may not occur during any operating time interval, modeling of physical and chemical processes may be approximate, and the quantities of the parameters of the models may not be precisely known. Quantifying the uncertainties in the context of a scenario-based risk model allows us to identify the aspects of the problem that are most important to risk. Therefore, characterizing all types of uncertainties is an essential element of risk assessment.

Measuring Knowledge

We have lower grades of knowledge, which we usually call degrees of belief, but they are really degrees of knowledge.... It may seem a strange thing to treat knowledge as a magnitude, in the same manner as length, weight, or surface. This is what all writers do who treat of probability, and what all their readers have done, long before they ever saw a book on the subject....By degree of probability we really mean, or ought to mean, degree of belief....Probability then, refers to and implies belief, more or less, and belief is but another name for imperfect knowledge, or it may be, expresses the mind in a state of imperfect knowledge.

—A. De Morgan (1847, *171–173*)

3.5.1 *Applying Bayes' Theorem*

Uncertainty is a probabilistic concept that is inversely proportional to the amount of knowledge, with less knowledge implying more uncertainty. *Bayes' theorem* is a common method of mathematically expressing a decrease in uncertainty gained by an increase in knowledge (for example, knowledge about failure frequency gained by in-field experience). Bayes' theorem has been particularly useful in quantifying the frequency of rare events (Apostolakis 1981).

Let λ_j be one failure rate of a set of possible failure rates of a component and E be a new body of evidence. We wish to express our knowledge of the probability of λ_j given E, which is expressed as $P(\lambda_j/E)$. Bayes' theorem gives us

$$P(\lambda_j / E) = \frac{P(\lambda_j)L(E/\lambda_j)}{\sum_j P(\lambda_j)P(E/\lambda_j)} \tag{3-1}$$

In summary this states that the knowledge of the "updated" probability of λ_j, given the new information E, equals the "prior" probability of λ_j before any new information times the likelihood function, $L(E/\lambda_j)$. The likelihood function is a probability that we really could have observed the new information, assuming that λ_j is the true failure rate. The numerator in Equation 3-1 is divided by a normalization factor, which must be such that the sum of the probabilities over the entire set of λ_j equals unity. Here is a sequence of steps in a typical application of Bayes' theorem within a PRA: (1) estimate the prior probability using engineering analysis, simulation, expert opinion, or any combination of the three; (2) obtain new information in the form of tests or experiments; (3) characterize the test information in the form of a likelihood function; and (4) perform the calculation in accordance with Equation 3-1 to infer the updated probability.

Both our prior and our updated state of knowledge can take on any functional form—Bayes' theorem is not restricted by the type of distribution. For aerospace, nuclear, and financial applications, we typically select the following: beta, binomial, Poisson, normal, lognormal, Rayleigh, Weibull, gamma, uniform, triangular, and noninformative.

3.5.2 *Eliciting and Structuring Expert Knowledge*

When we apply Bayes' theorem in PRA, we often must construct a prior distribution for which actual experiential data are rare. In this case, we need the knowledge or judgment of experts. A common use of engineering judgment is predicated on the reasonable notions that sometimes the data reside only within an expert's mind or that the data can be reasonably interpreted and used only through an expert's background and experience. The elicitation of knowledge, including but not limited to probabilities, is a large field of study and practice by itself. The literature contains books, reports, and papers from the fields of psychology, management science, decision theory, probability and statistics, and nuclear engineering (see, for example, Shooman and Sinkar 1977; Kahneman et al. 1982; Mosleh et al. 1987; Apostolakis 1988; Cooke 1991; Tregoning et al. 2005). Although methods in the literature differ, we can summarize a general procedure for a comprehensive elicitation of knowledge in four steps:

1. Preparation by the risk analysts
2. Knowledge elicitation session with the experts
3. Calculation of scores and
4. Combination of judgments

Although analysts can use a variety of specific techniques and analyses at each step, we can generally describe them as follows:

1. *Preparation by the risk analysts* includes identifying parameters that are significant potential sources of uncertainty, developing questions for the experts, and choosing a moderator for the elicitation session. If the elicitation is designed to help develop a model as well as estimate parameter quantities, steps 1 and 2 are conducted iteratively.
2. *Knowledge elicitation session with the experts* can be performed in five steps:
   a. Motivate the expert (explain the reason for conducting the session).
   b. Structure the discussion (tell the expert exactly what is expected and in what format).
   c. Precondition the expert to reveal biases and encourage truthful judgments.
   d. Encode the information (record what the expert says).
   e. Play back the information to get the expert's confirmation.

After the elicitation session, the risk analyst evaluates the answers to make sure that the responses truly reflected what was required of the model. Steps 2 and 3 may require iteration as well.

3. *Calculation of scores* refers to assigning a numerical metric to an expert's probability. Scoring is used to reward (or condition) an expert's opinion. In addition, the scores can evolve into weights for combining probability assessments from different experts. Developing a scoring method involves two questions (Cooke 1991):
   a. Does the score reward those features that are desired in subjective probability assessments?
   b. Does the score introduce a reward structure that distorts or biases the assessment of probabilities?
4. *Combination of judgments*. PRA practitioners have devoted much effort to combining the judgment of multiple experts who are rendering an opinion on the same problem. For example, seismic risk analyses of nuclear power

units rely heavily on the use of multiple experts to estimate the seismicity of a region and the strength of equipment and structures. Similarly, multiple experts are involved in the estimation of source terms in hypothesized nuclear power plant accident scenarios. Cooke (1991) divides the range of models for combining expert judgments into three categories:

a. *Classical.* This constructs a weighted combination of expert probability assessments.

b. *Bayesian.* These methods take the perspective of a DM. A Bayesian prior distribution is modified using opinions developed by experts, taking into account the DM's attitude toward each expert's opinion.

c. *Psychological scaling.* These models for estimating probability distributions originally derive from the estimation of relative intensities of psychological stimuli based on pairwise comparison. They do not produce numerical estimates. Instead, they provide relative scales of the expert opinions with one or more degrees of freedom used for independent calibration.

3.5.3 *Propagating Uncertainties through a Risk Model*

Representing uncertainties as probability distributions is fundamental to risk and risk assessment. Uncertainties can be propagated through a risk model by means of Monte Carlo simulation. Alternative forms of sampling that optimize the sampling space, such as Latin Hypercube, are also used in risk assessment.

The concept of Boolean reduction and solution of fault trees is introduced in Sections 3.3.4 through 3.3.7. In Figure 3-15, minimal cutsets are shown as the products of the basic event probability distributions, b_{ij}, in which i refers to the cutset and j refers to a basic event within the cutset. We can perform multiplication and summation of the distributions using a Monte Carlo simulation technique that I explain next. If we use an FT to disaggregate an initiating or pivotal event, this top event probability distribution becomes the initiating or pivotal event probability distribution. Figure 3-16 illustrates propagation of the distributions through an ET. Each scenario ending in the end state "Bad" is quantified by probabilistic multiplication, using Monte Carlo simulation or another sampling technique. The total probability of ending in a Bad end state is the probabilistic sum of all the probability distributions of scenarios that end in Bad. This is shown in Figure 3-16.

Monte Carlo simulation allows operation with probability distributions for any algebraic equation as if the operations were performed with simple numbers. This is because the Monte Carlo method simulates operations on probability distributions by picking random numbers from each distribution and using these in the algebraic equation. Each selection of random numbers and combination via the equation constitutes one Monte Carlo trial. The resultant distribution is better approximated by increasing the number of trials. The mean error is inversely proportional to the square root of the number of trials.

Figure 3-17 illustrates how a Monte Carlo simulation achieves this combination of distributions. The example in the figure adds two uncorrelated distributions, P(x) and P(y).

First, we convert all distributions into cumulative density functions (CDFs) as the figure shows. Next, we perform many Monte Carlo trials. In each, we carry out the following tasks:

1. Pick a random number between 0 and 1. This corresponds to randomly selecting a confidence level along the y-axis of P(x).
2. Find the corresponding x and call it X_1.
3. Pick another random number between 0 and 1. This corresponds to randomly selecting a confidence level along the y-axis of P(y) in Figure 3-17. (We choose two separate random numbers when the distributions are uncorrelated. If they are completely correlated, we use the same random number for both distributions.)
4. Find the corresponding y and call it Y_1.
5. Add $X_1 + Y_1 = C_1$.
6. Steps 1 through 5 constitute one Monte Carlo trial. Repeat steps 1 through 5 such that the outcome of each is $X_i + Y_i = C_i$.
7. Bin or sort the trials as described in Figure 3-18.

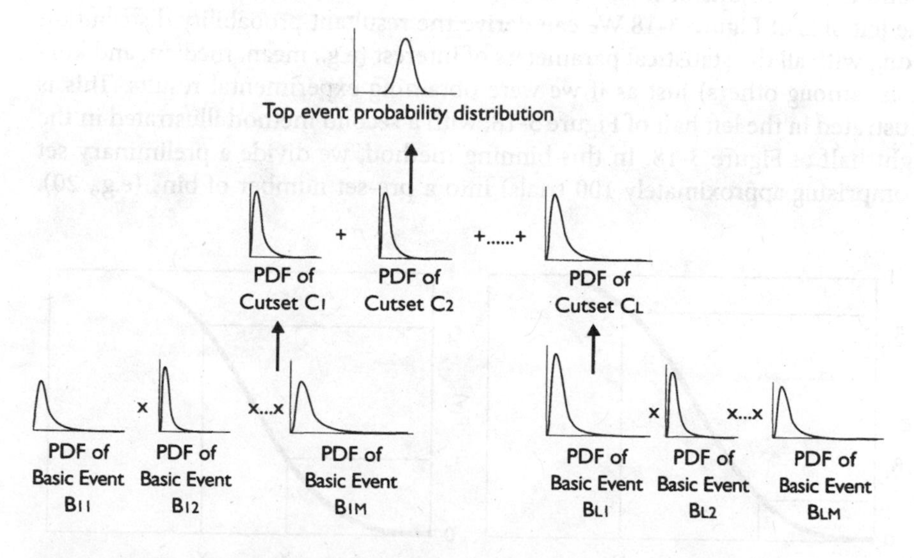

Figure 3-15. ***Propagation of Probability Distributions through a Fault Tree***

Note: PDF = probability density function.

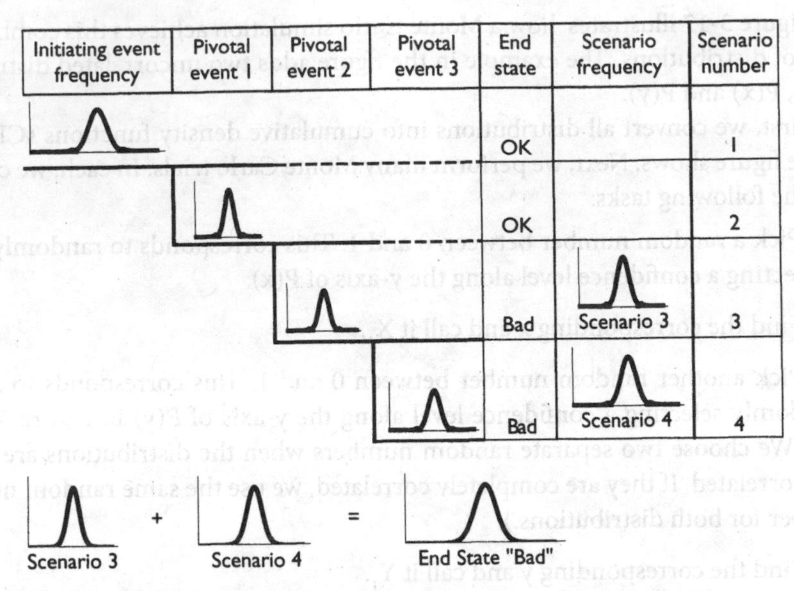

Figure 3-16. ***Propagation of Uncertainties through an Event Tree***

One of the fundamental principles of this method is that each trial is equally likely—therefore, the number of trials that result in a particular result, C, is proportional to the probability of its occurrence. In the simplest of Monte Carlo methods, the results of many trials are sorted in ascending order as shown in the left side of Figure 3-18.We can derive the resultant probability distribution along with all the statistical parameters of interest (e.g., mean, median, and kurtosis, among others) just as if we were obtaining experimental results. This is illustrated in the left half of Figure 3-18, with a second method illustrated in the right half of Figure 3-18. In this binning method, we divide a preliminary set (comprising approximately 100 trials) into a pre-set number of bins (e.g., 20).

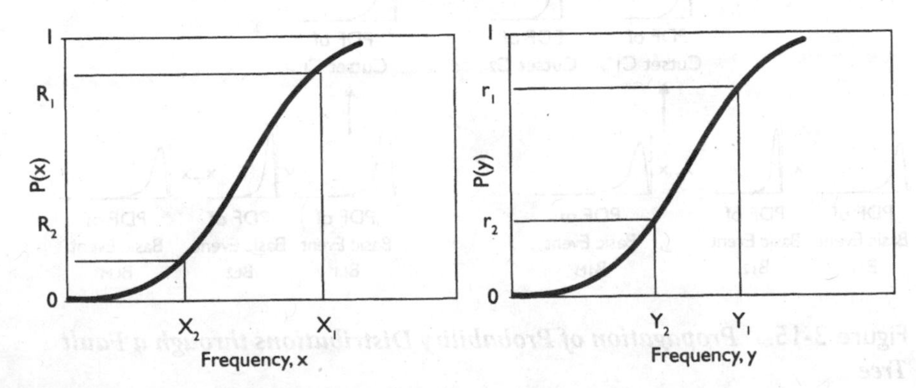

Figure 3-17. ***Monte Carlo Addition***

We place subsequent trials in the appropriate bin. After all trials are complete, we tally the number of trials in each bin. The probability/confidence intervals of each bin is proportional to the number of trials in that bin.

3.6 Presenting Results

For events or end states, such as loss of vehicle, core damage, or mission failure, a probability distribution over frequency of such events is the generally accepted format for presenting the results. (Refer to my discussion of probability distributions in Section 2.2.) Such distributions take a variety of tabular and graphical forms as depicted in Figures 3-19 and 3-20 and in Table 3-1. The frequency of the

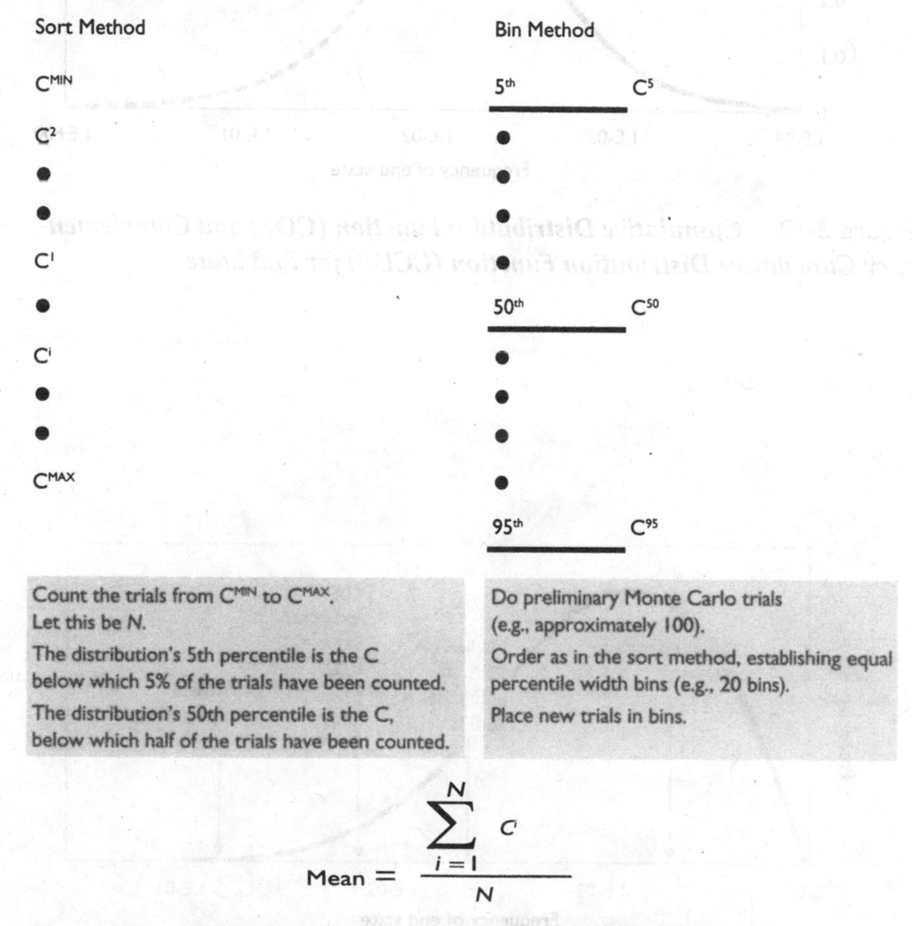

$$\text{Mean} = \frac{\sum_{i=1}^{N} C^i}{N}$$

Figure 3-18. *Sort and Bin Construction of Probability Distribution from Monte Carlo Simulation*

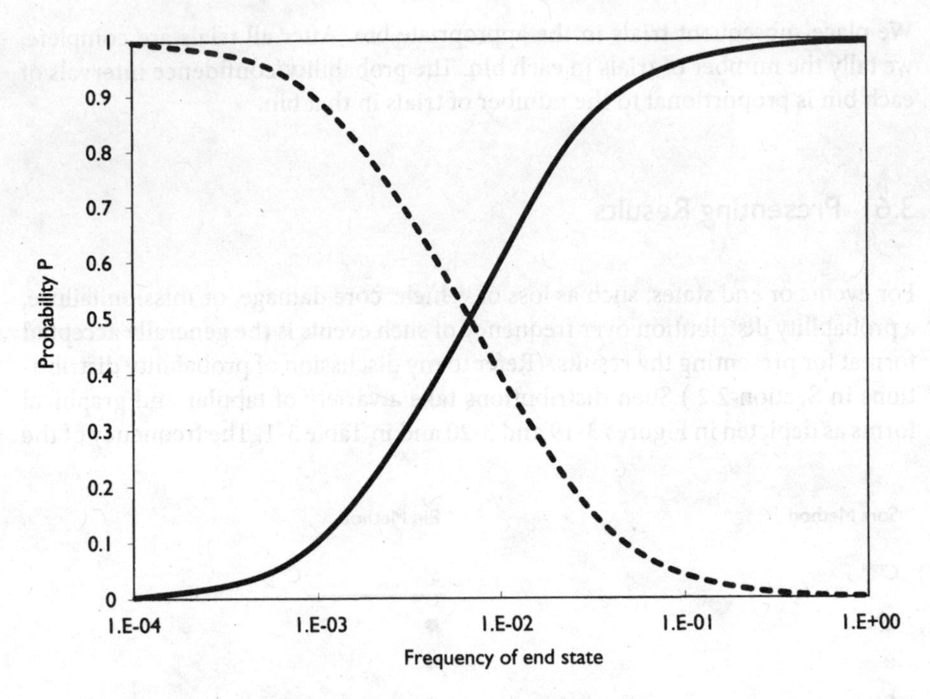

Figure 3-19. *Cumulative Distribution Function (CDF) and Complementary Cumulative Distribution Function (CCDF) for End State*

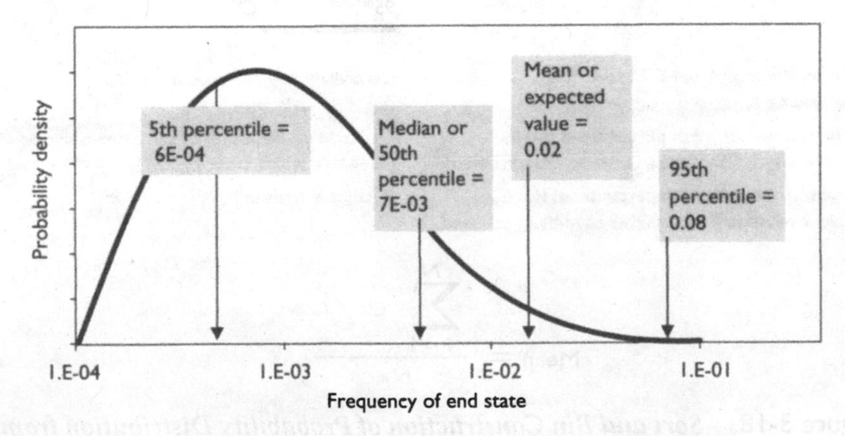

Figure 3-20. *Probability Density Function (PDF) for End State*

end state (x-axis in the figure) can be, for example, the frequency of failure of a launch vehicle to reach orbit (per mission); the frequency of an aircraft accident (per flight-hour or per mile traveled); the frequency of core damage in a nuclear reactor (per year); or the frequency of catastrophic fire (per hour of operation).

Note that Figure 3-19 illustrates a CDF and a CCDF. A CDF represents the estimated probability (on the y-axis) that the frequency of the end state is *at or below* the corresponding quantity on the x-axis. A CCDF, which is the probability complement of a CDF, represents the estimated probability (on the y-axis) that the frequency of the end state *is at or above* the corresponding quantity on the x-axis. The PDF, which is the mathematical derivative of the CDF (see Figure 3-20) may be a more familiar way to present a probability distribution. Recall that in a PDF, probability is represented by the area under the curve. The median or 50th percentile is the frequency at which there is an equal area on the left and on the right. We do not see this immediately because the figure is logarithmic on the x-axis. The 95th percentile is the frequency at which 95% of the area is to the left, which indicates a 95% probability that the frequency is equal to or less than the 95th percentile (on the x-axis). Similarly, the 5th percentile is the frequency at which only 5% of the area is to the left, indicating a 5% probability that the frequency is equal to or less than the 5th percentile on the x-axis. We often give the 95th percentile as a reasonable upper bound of the result and the 5th percentile as a reasonable lower bound of the result. Typically, we give the median and mean as measures of central tendency or best estimate. The mean is sometimes referred to as expected value. And sometimes, we simply present the information in a table, such as in Table 3-1.

For end states characterized by a range of numerical quantities, such as a number of injuries, an amount of financial loss in dollars, or an amount of time lost because of injuries, we express results in terms of a family of either CDFs or CCDFs. Figure 3-21 shows an example from a study of accidents that might occur to a launch vehicle carrying a plutonium-fueled, radioisotope, thermal electric generator.

Each curve is one instantiation of the results of the risk assessment. Each curve in a family of CCDFs conveys a level of confidence that the true frequency/consequence relationship is on or below the curve. The 5% confidence risk curve is located such that there is only a 5% probability (according to the analysis) that the actual results lie on or below that curve. Similarly, the 95% confidence risk curve is located such that there is 95% probability that the actual results lie on or

Table 3-1. ***Tabularized Probability Distribution Results***

Parameter	*Frequency of End State*
Fifth percentile	6E-04
Median	7E-03
Mean	0.02
Ninety-fifth percentile	0.08

below that curve. The 50% confidence risk curve is located such that there is an equal chance that the results fall above or below it. The mean curve is that curve used for decisionmaking purposes.

Here are some other examples of result presentations of risk assessments.

U.S. Space Shuttle APU. Prior to 1987, analysts had performed a large failure mode and effects analysis (per Mil-Std-1629) and a hazard analysis (per Mil-Hdbk-882) on the space shuttle's APU. The analyses revealed hundreds of failure modes and numerous failure scenarios. Even though engineers attempted to mark each failure mode or hazard as critical or noncritical, they could not be clearly ranked in terms of their importance to overall shuttle risk because the analysts did not use quantitative risk analysis methods. Without such a risk-based ranking, DMs with limited budgets did not know the best way to spend money to reduce risk.

In 1987, a comprehensive risk assessment was performed (Frank 1989). This analysis was rigorously quantitative, including quantitative characterization of uncertainties. The results, however, could be displayed very simply. As shown in Figure 3-22, 99% of the risk came from only 20 of the more than 300 failure modes. This significant finding was meant to help shuttle managers decide how to improve the APUs.

International Space Station. Instead of showing probability distributions for end states (as in Figures 3-19 and 3-20), a PRA of the International Space Station presented probability bands. This is illustrated in Figure 3-23.

The left end of each band is the 5th percentile, and the right end is the 95th percentile. A vertical line could be added to represent the mean.

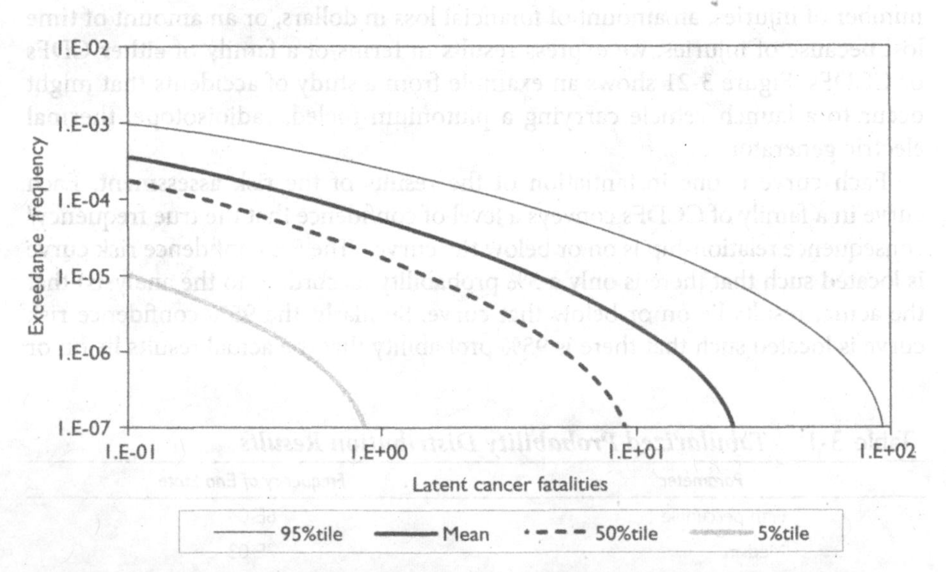

Figure 3-21. ***Family of Complementary Cumulative Distribution Functions for Varying Numerical Consequences***

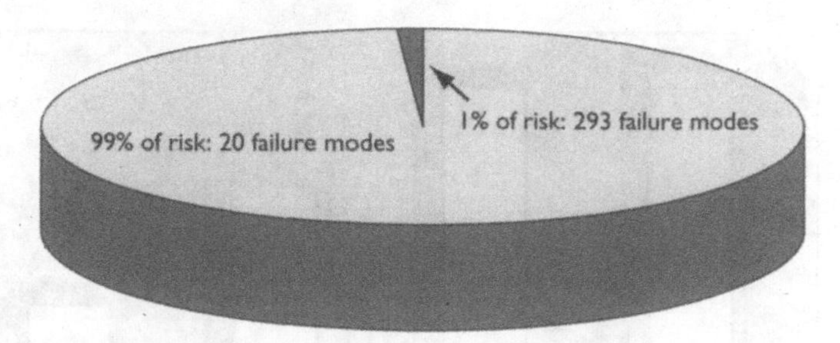

Figure 3-22. ***Risk Associated with Critical and Noncritical Items of Shuttle Auxiliary Power Units***

Mars Mission Architecture. In this example, NASA mission architects were attempting to design a mission to map the climate of Mars. This mapping was to be done by casting numerous landers across the planet's surface. The architects wanted to know how many landers would be required to be launched toward Mars to ensure a 90% confidence of success if 12 working spacecraft were required for mission success. Figure 3-24 clearly shows the results of a sensitivity study in which the number of launched landers was varied. (See Chapter 9, where I present more detail about this study and the results.) Notice that the bars decrease in height as the probability of mission failure decreases. If 12 spacecraft are required for success, at least 20 must be launched to ensure less than a 10% chance of failure (as shown in the figure on the cross-hatched option bar over "12 of 20"). This sensitivity study resulted in important guidance to the mission architects with respect to cost, feasibility, weight, and other factors.

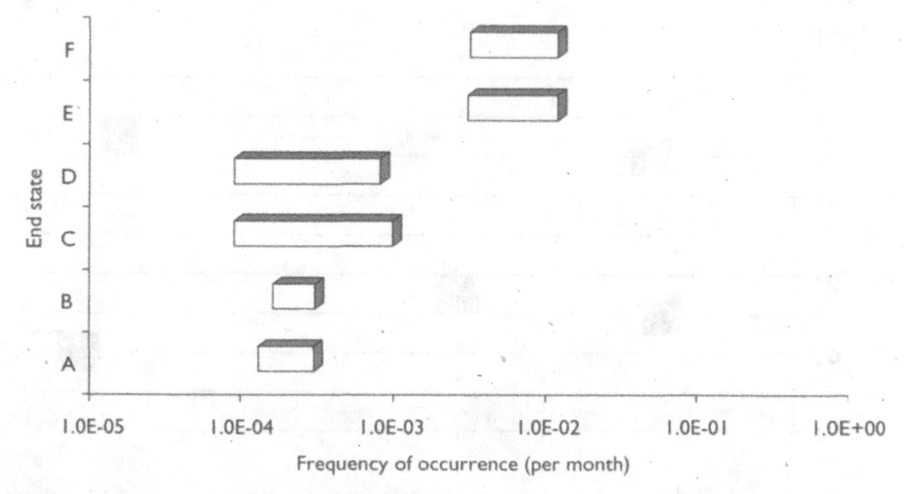

Figure 3-23. ***Summary of International Space Station Probabilistic Risk Analysis Results***

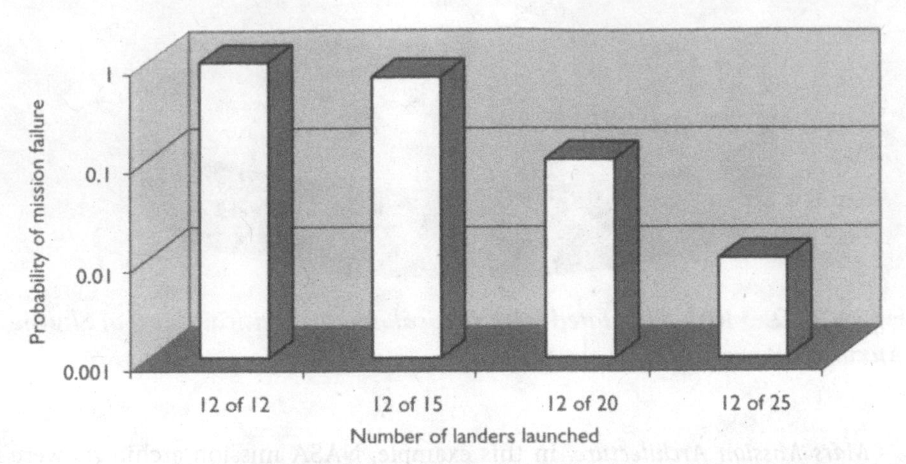

Figure 3-24. ***Sensitivity of Mission Failure to Number of Launched Landers***

Automotive Development Program Risk. DMs at an automotive company wanted to track development risks of a new product. They identified technology risks, part acquisition risks, and product schedule risks. All were given an effective cost. The DMs elected to show their risks in the form of a probability-versus-cost graph, such as Figure 3-25. Each risk (in a box) has a probability of occurrence and a cost penalty during a project. During the course of the project, the risks were tracked and the graph modified as shown, for example, in Figure 3-26.

Permit Application Risk. Project DM's wanted to track the financial risks of developing, tendering, and receiving an approval to construct and operate a new electrical power station. They reasoned that the risks were largest at the

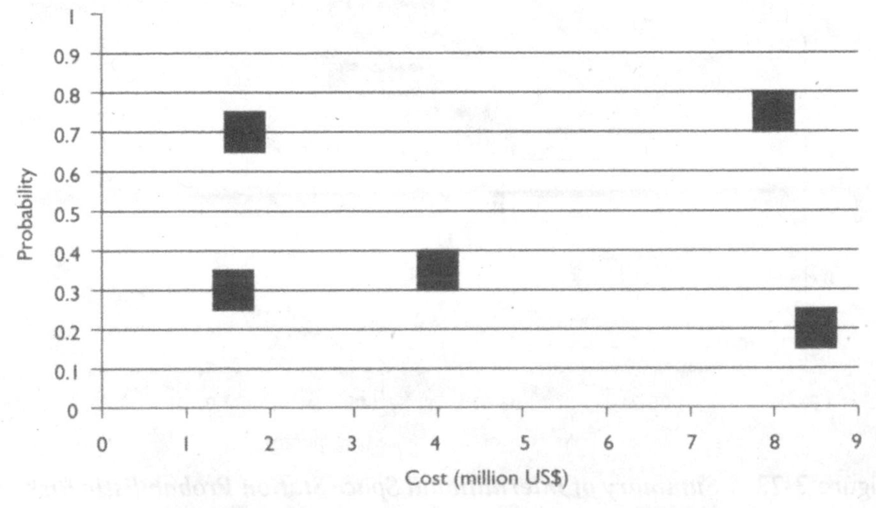

Figure 3-25. ***Risk Management Graph at Beginning of Project***

beginning of the project and would ideally decrease as the power plant design and permit applications were refined. They periodically identified programmatic risks and updated the cost estimates of these risks to identify deviations from a decreasing trend. They plotted total project risk as a function of schedule time (see Figure 3-27). All risks were associated with cost so that the DMs obtained the total project risk as the probabilistic sum of the individual risks. The DMs used this picture to actively implement risk reduction strategies to achieve a decreasing trend.

3.7 Survey of Other PRA Methods and Techniques

Up to this point in this chapter, I have described the core methods most often used in PRA. In this section, I briefly survey a selection of other methods.

3.7.1 Use of Simulation and Other Dynamic Methods

Simulation is a completely different paradigm for risk assessment than the ETs, FTs, and MLDs I described in Section 3.3, which comprise a family of diagrams sometimes called *belief networks.* In the belief network paradigm, the analyst thinks through a system and develops diagrams that explicitly depict scenarios. In simulation methods, the analyst develops a mathematical model that must be coded as software. A perturbation is introduced to the system (i.e., an initiating

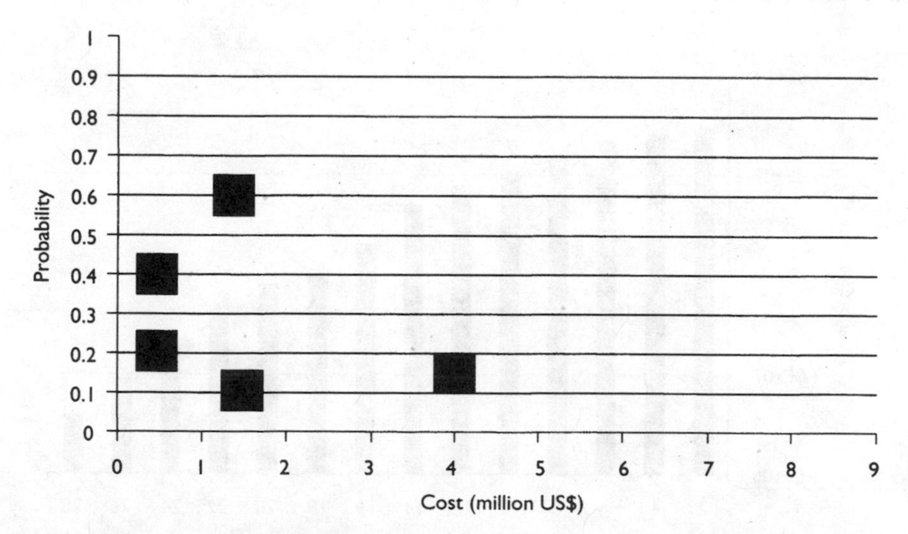

Figure 3-26. ***Risk Management Graph at Middle of Project after Implementing Risk Reduction Strategies***

event), and the effect of that perturbation (i.e., an end state) results from the simulation.

Variations in the perturbations, variations in the response of the system, and uncertainty in knowledge about the system give rise to alternative scenarios. Because some processes are inherently stochastic or chaotic, we also see variations in system response. For example, risk assessments often involve analyzing energetic reactions to an initiating event, such as a fire or explosion. Furthermore, risk assessment can involve dependencies with respect to the timing of events. Finally, to assess the consequences of scenarios—such as, for example, the amount of toxic material release and the subsequent number of injuries—we must often make probabilistic computations of the governing physical, chemical, and biological processes. When we use a simulation method, our objective is to characterize the likelihood of alternative scenarios and the probabilities of end states by simulating the situation many times (called *trials*). We use sampling methods to meet this objective. The two most commonly used in PRA are Monte Carlo and Latin Hypercube.

Over the last decade, simulation methods have been most widely applied to nuclear power risk assessments in space (e.g., Lockheed Martin 1997). In this application, analysts consider scenarios to be dynamic in that the response to an initial perturbation evolves over time because of component, phenomenological, and environmental interactions. For example, the potential for and amount of radioactive material released depends on the magnitude of previous explosions, fragment fields, and the altitude of the vehicle. The fragment field, in turn, depends on the explosion's magnitude and the vehicle's altitude. Furthermore, explosions are inherently uncertain phenomena. Even slight differences in conditions can cause differences in explosion overpressures and fragment velocities

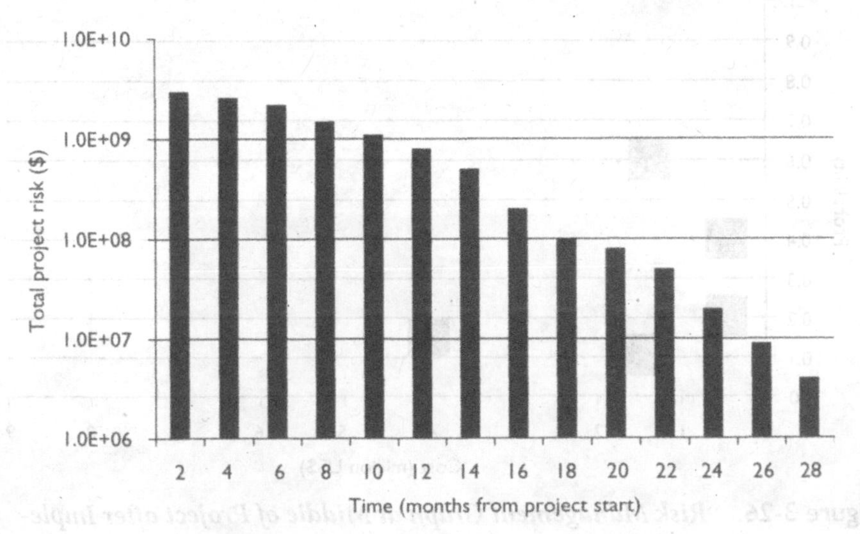

Figure 3-27. *Tracking Total Project Risk*

that change the probability of radioactive material release. In this example, the analyst models the entire set of physical, chemical, and biological processes at work. The analysis yields probability distributions over such quantities as the amount of released material and the number of latent cancer fatalities.

Although simulation methods offer a powerful technique for modeling dynamic systems, the ability to clearly visualize scenarios, the relationship of end states to initiating events, and the associated phenomena is lost. ETs and FTs, on the other hand, allow representational clarity but may not be as accurate for dynamic systems with many interrelated processes.

Other dynamic methods are the dynamic event tree method (Acosta and Sui 1993) and the dynamic flowgraph method (Milici et al. 1996). These methods have similarities to both belief networks and simulation methods. Their usefulness arises from the ability to model dynamic scenarios, especially involving human actions, while maintaining some representational clarity. Siu (1994) and Devooght and Smidts (1993) present surveys of dynamic methods.

The dynamic ET approach is characterized by branching at discrete time intervals during an accident progression. It involves applying sets of information or rules for (1) defining a branching set, (2) branching to create different scenarios, (3) defining system states, (4) expanding the scenarios, and (5) quantifying the scenarios (Acosta and Sui 1993). The solution algorithm tracks forward in time.

The dynamic flowgraph method starts with a directed graph of the system. As in FTs, the algorithm develops prime implicants of the system. It does this by parsing backward in time through the directed graph by obtaining the prime implicants for the most recent time and then backtracking in time to discover how these prime implicants were caused (Milici et al. 1996).

These rules and algorithms are similar, in concept, to techniques used in object-oriented programming, in which objects and their interrelationships behave according to well-defined rules (such as equations-of-state). For realistic problems, both the simulation and the dynamic tree methods tend to be labor and computationally intensive. In complicated systems for which dynamic interactions are crucial for understanding risk, though, we can obtain useful results from both methods (Siu 1994).

For some physical, chemical, and biological process analyses, we do not need simulation or dynamic trees because the computations are either solvable in closed form or well-known numerical solutions are available. I describe the theory and use of probabilistic physical process modeling within the belief net paradigm for aerospace mission risk in Frank (1999).

3.7.2 *Software Risk*

Although software reliability methods are reasonably well established (e.g., Shooman 1983; Musa et al. 1990), software risk is an emerging area of PRA. Software reliability is useful during a software development process in which the software developers attempt to predict when the development has reached an acceptable level of residual errors so that it can be released for open use. The operational software failure rate is often taken as the failure rate prediction from the methods at the time of release. Analysts have used several methods (ranging from FTs to formal methods) to attempt to demonstrate that a particular software function has been adequately designed (e.g., Parnas et al. 1990; Bowman and Leveson 1991; Leveson 1994).

The main impetus for software risk assessment, however, is the recognition that we are increasingly using digital-control systems that are *driven* by software to control safety related systems (Dunn et al. 1994). In software risk assessment, we take on the more difficult task of identifying specific software-failure modes within the context of scenarios. To do this, we must investigate the software within the context of the electrical, mechanical, structural, environmental, human, and electronic systems within which it operates. The objective is to include software errors and their likelihood within the overall set of accident scenarios.

Software errors acting alone or in combination with hardware errors can be contributing causes of initiating events or pivotal events in an accident scenario. Failure to properly sequence the starting of the space shuttle main engines (SSMEs) and the firing of solid rocket motors prior to liftoff are examples of initiating events that can result from one or more software errors. An example of a software-related pivotal event is failure to command an SSME shutdown once a specified extreme limit (called a "redline") has been exceeded. Within the context of a PRA, the area of software risk is at an early stage of development.

3.7.3 Human Reliability Analysis

In this area, the technology has its genesis in the defense and nuclear industries (Swain and Guttman 1983; Hannaman and Spurgin 1984). Documented mishaps in all industries as well as risk assessment predictions demonstrate the importance of human actions in accident scenarios (e.g., Reason 1997). Researchers around the world have developed an enormous body of research and a number of methods designed to understand the ability of operators to respond to and recover from nuclear accident scenarios. Several survey books that summarize the field up to the time of their publication are available (e.g., Dougherty and Fragola 1988; Hollnagel 1993; Gertman and Blackman 1994). Each chooses to characterize and categorize methods in a different way. We can distinguish, however, a first generation of methods (e.g., Hall et al. 1982; Swain and Guttman 1983; Embrey et al. 1984; Hannaman and Spurgin 1984) from a second generation of methods (e.g., Spurgin and Moieni 1991; Hollnagel 1998; U.S. NRC 1999). The first generation of methods consider only phenotypes; in other words, only the outcome or the observed manifestation of human endeavor are modeled. The second generation attempts to account for both phenotype and genotype, which is the cause—involving human cognitive function—underlying the specific action. The first generation of methods is geared toward "filling in the blanks" in an FT or an ET that contains a box for a human error probability in a manner analogous to hardware failures. The second generation takes a more sophisticated approach to modeling human actions that considers the context and environment within which people are working.

3.7.4 *External Event Risk*

External events are initiating events whose genesis is external to the system in question. In aircraft and space applications, for example, these are typically wind, lightning, bird strikes, external atmospheric and orbital debris, and micrometeoroid impacts. For tank farms or nuclear power plants, earthquakes, flooding, high winds, and lightning strikes are external factors that are significant in terms of risk. By their very nature, these initiating events can disable multiple components and multiple systems. For well-designed redundant systems, the conditional probability of catastrophic failure in the face of an external, high-energy initiating event is higher than that from a nonenergetic component or a system failure. In analyzing such energetic failures, a key characteristic is that the risk depends on the spatial proximity of other components and systems. A risk assessment analyst must keep track of two fundamental parameters—the capacity of the energetic event to spread to adjacent components or areas, and a spatial inventory of equipment in these areas.

Notes

1. In PRA we deal with the problem of many paths in various ways—such as event trees, simulations, and probability distributions—as I discuss later in this chapter.
2. In Sections 3.3 through 3.5, I summarize a large body of knowledge about PRA. More details on the structure and methods of PRA can be found in, for example, ANS and IEEE (1983); Modarres (1992); Kumamoto and Henley (1996); and Bedford and Cooke (2001). Leveson (1995) is a good reference for a nonprobabilistic view of system safety with and without software.
3. See Section 3.3.4 for definitions of gates.
4. Refer to Figure 2-2 for the valve configuration.
5. The numbers in this example should be taken as fictitious.
6. This becomes a trade-off in design and operation for systems that are not normally in operation (e.g., on standby status). For such systems, preventive maintenance tends to increase operational availability of each component but also increases the fraction of such component unavailability episodes in which the entire system is down. This is largely because of human errors associated with performance of the maintenance tasks.

References

Acosta, C., and N. Sui. 1993. Dynamic Event Trees in Accident Sequence Analysis: Application to Steam Generator Tube Rupture. *Reliability Engineering and System Safety* 41(2): 135–154.

ANS (American Nuclear Society) and IEEE (Institute of Electrical and Electronics Engineers). 1983. PRA Procedures Guide. NUREG/CR-2300. Washington, DC: U.S. Nuclear Regulatory Commission (U.S. NRC).

Apostolakis, G. 1981. Bayesian Methods in Risk Assessment. In *Advances in Nuclear Science and Technology*. Edited by J. Lewins and M. Becker. New York: Plenum Press.

———. 1988. *Expert Judgment in Probabilistic Safety Assessment, Accelerated Life Testing and Expert's Opinions in Reliability*. Edited by C.A. Clarotti and D.V. Lindley. Corso, Italy: Società Italiana di Fisica.

Bedford, T., and R. Cooke. 2001. *Probabilistic Risk Assessment: Foundations and Methods*. Cambridge, UK: Cambridge University Press.

Bowman, W.C., and N. Leveson. 1991. An Application of Fault Tree Analysis to Safety Critical Software at Ontario Hydro. In *Probabilistic Safety Assessment and Management*. Edited by G. Apostolakis. New York: Elsevier, 363–368.

Cooke, R.M. 1991. *Experts in Uncertainty: Opinion and Subjective Probability in Science*. Oxford, UK: Oxford University Press.

De Morgan, A. 1847. *Formal Logic*. London: Taylor and Walton.

Denson, W., G. Chandler, W. Crowell, A. Clark, and P. Jaworski. 1995. *Non-Electronic Part Reliability Data 1995*. Rome, NY: Reliability Analysis Center (RAC).

Denson, W., W. Crowell, P. Jaworski, and D. Mahar. 1997a. *Failure Mode/Mechanism Distributions 1997*. Rome, NY: RAC.

———. 1997b. *Electronic Parts Reliability Data: A Compendium of Commercial and Military Device Field Failure Rates*. Rome, NY: RAC.

Devooght, J., and C. Smidts. 1993. Probabilistic Dynamics: The Mathematical and Computing Problems Ahead. In *Reliability and Safety Assessment of Dynamic Process Systems*. Edited by T. Aldemir, N. Sui, A. Mosleh, P.C. Cacciabue, and G. Goktepe. New York: Springer-Verlag.

DoD (U.S. Department of Defense). 1991. *Military Handbook Reliability Prediction of Electronic Equipment*. Mil-Hdbk-217F. Washington, DC: DoD.

Dougherty, E.M., Jr. and J.R. Fragola. 1988. *Human Reliability Analysis*. New York: John Wiley and Sons.

Dunn, W., M. Frank, S.A. Epstein, and L. Doty. 1994. Risk Assessment and Management of Safety-Critical, Digital Industrial Controls—Present Practices and Future Challenges. In *Proceedings of International Conference on Probabilistic Safety Assessment & Management (PSAM-II)*. March 20–25, San Diego, California.

Embrey, D.E., P. Humphrys, E.A. Rosa, B. Kirwan, and K. Rea.1984. *SLIM-MAUD. An Approach to Assessing Human Error Probabilities Using Structured Expert Judgment*. NUREG/CR-3518. Washington, DC: U.S. NRC.

Fleming, K.N. 1975. A Reliability Model for Common Mode Failures in Redundant Safety Systems. In *Proceedings of the Sixth Annual Conference on Modeling and Simulation*.

Fleming, K.N., and A.M. Kalinowski. 1983. *An Extension of the Beta Factor Method for Systems with High Levels of Redundancy*. PLG-0289. Newport Beach, CA: PLG.

Fleming, K.N., and S.B. Rao. 1992. *A Database of Common Cause Events for Risk and Reliability Applications*. EPRI TR-100382. Palo Alto, CA: Electric Power Research Institute (EPRI).

Frank, M. 1989. Quantitative Risk Assessment of a Space Shuttle Subsystem. In *Proceedings of PSA '89 International Topical Meeting—Probability, Reliability, and Safety Assessment*. Washington DC: American Nuclear Society.

———. 1999. Treatment of Uncertainties in Space Nuclear Risk Assessment with Examples from Cassini Mission Applications. *Reliability Engineering and System Safety* 66: 203–221.

Frank, M., M. Weaver, and R. Baker. 2005. A Probabilistic Paradigm for Spacecraft Random Reentry Disassembly. *Reliability Engineering and System Safety* 90(2–3): 148-161.

Gertman, D., and H. Blackman. 1994. *Human Reliability and Safety Analysis Data Handbook*. New York: John Wiley & Sons.

Hall, R.E., J. Fragola, and J. Wreathall.1982. *Post-event Human Decision Errors: Operator Action Trees/Time Reliability Correlations*. NUREG/CR-3010. Washington, DC: U.S. NRC.

Hannaman, G., and A. Spurgin. 1984. *Systematic Human Action Reliability Procedure (SHARP)*. EPRI NP-3583. Palo Alto, CA: EPRI.

Hollnagel, E. 1993. *Human Reliability Analysis: Context and Control*. London: Academic Press.

———. 1998. *Cognitive Reliability and Error Analysis Method: CREAM*. London: Elsevier Science.

Kahneman, D., P. Slovic, and A. Tversky (eds.). 1982. *Judgment under Uncertainty: Heuristics and Biases*. New York: Cambridge University Press.

Kumamoto, H., and E.J. Henley. 1996. *Probabilistic Risk Assessment and Management for Engineers and Scientists*, 2nd ed. Piscataway, NJ: IEEE Press.

Leveson, N. 1994. Software Safety Analysis. In *Proceedings of the High Consequence Operations Safety Symposium*. SAND94-2364. Albuquerque, NM: Sandia National Laboratory.

———. 1995. *Safeware: System Safety and Computers*. New York: Addison-Wesley.

Lockheed Martin. 1997. *General Purpose Heat Source-Radioisotope Thermoelectric Generators in Support of the* Cassini *Mission Final Safety Analysis Report (FSAR)*. CDRL C.3. Valley Forge, PA: Lockheed Martin Missiles and Space, Valley Forge Operations.

Marshall, F.M., A. Mosleh, and D.M. Rasmussen.1998a. *Common Cause Failure Database and Analysis System, Volumes 1–4*. NUREG/CR-6268. Washington, DC: U.S. NRC.

———. 1998b. *Common Cause Failure Parameter Estimations*. NUREG/CR-5497. Washington, DC: U.S. NRC.

Milici, A., J.-S. Wu, and G. Apostolakis. 1996. The Use of the Dynamic Flowgraph Methodology in Modeling Human Performance and Team Effects. In *Probabilistic Safety Assessment and Management*. Volume 1. New York: Springer-Verlag.

Modarres, M. 1992. *What Every Engineer Should Know about Reliability and Risk Assessment*. Boca Raton, FL: CRC Press.

Mosleh, A., V.M. Bier, and G. Apostolakis. 1987. *Methods for the Elicitation and Use of Expert Opinion in Risk Assessment*. NUREG/CR-4962. Washington, DC: U.S. NRC.

Mosleh, A., K. Fleming, and G. Parry, et al.1988. *Procedures for Treating Common Cause Failures in Safety and Reliability Studies*. NUREG/CR-4780. Washington, DC: U.S. NRC.

Mosleh, A., D.M. Rasmuson, and F.M. Marshall.1998. *Guidelines on Modeling Common Cause Failures in Probabilistic Risk Assessment*. NUREG/CR-5485. Washington, DC: U.S. NRC.

Musa, J., A. Iannino, and K. Okumoto.1990. *Software Reliability: Measurement, Prediction, Application*. New York: McGraw Hill.

NSWC (Naval Surface Warfare Center). 1998. *Handbook of Reliability Prediction Procedures for Mechanical Equipment*. NSWC LE1. West Bethesda, MD: Naval Surface Warfare Center, Carderock Division.

Parnas, D.L., A. J. van Schouwen, and S.P. Kwan. 1990. Evaluation of Safety Critical Software. *Communications of the Association of Computing Machinery* 33(6): 636–648.

Rasmuson, D.M., A. Mosley, and F. Marshall. 1998. Some General Insights from the USNRC's Common Cause Database. In *Probabilistic Safety Assessment and Management: PSAM 4*. Edited by A. Mosley and R.A. Bari. New York: Springer-Verlag, 195–199.

Reason, J. 1997. *Managing the Risks of Organizational Accidents*. Hampshire, UK: Ashgate Publishing.

D.I. Gertman, W.E. Gilmore, W.J. Galyean, et al. 1988. *Nuclear Computerized Library for Assessing Reactor Reliability (NUCLARR)*. NUREG/CR-4639. Washington, DC: U.S. NRC.

Shooman, M. 1983. *Software Engineering*. New York: McGraw Hill.

Shooman, M.L., and S. Sinkar. 1977. Generation of Reliability and Safety Data by Analysis of Expert Opinion. In *Annual Reliability and Maintainability Symposium*. New York: IEEE, 186–193.

Siu, N. 1994. Risk Assessment for Dynamic Systems: An Overview. *Reliability Engineering and System Safety* 43: 43–73.

Spurgin, A., and P. Moieni. 1991. Interpretation of Simulator Data in the Context of Human Reliability Modeling. In *Probabilistic Safety Assessment and Management*. Edited by G. Apostolakis. New York: Elsevier.

Swain, A., and H. Guttman. 1982. *Handbook of Human Reliability Analysis with Emphasis on Nuclear Power Plant Applications*. NUREG/CR-1278. Washington, DC: U.S. NRC.

Telcordia Technologies. 2001. *Reliability Prediction Procedure for Electronic Equipment*. SR-332, Issue 1, May. Piscataway, NJ: Telcordia.

Tregoning, R., L. Abramson, and P. Scott. 2005. *Estimating Loss-of-Coolant Accident (LOCA) Frequencies through the Elicitation Process*. NUREG-1829. Washington, DC: U.S. NRC.

U.S. NRC. 1999. *Technical Basis and Implementation Guidelines for a Technique for Human Event Analysis (ATHEANA)*. NUREG-1624, Revision 1. Washington, DC: U.S. NRC.

CHAPTER 4

Procedures for Making Safety-Related Decisions

A DECISION IS A CHOICE among alternative courses of action (or options) made to meet an overall objective. Decisions that involve the safety of complicated equipment rarely involve a single consideration. In other words, a decisionmaker (DM) will mull over several questions about improving safety: What is the cost to the project? What is the sacrifice in reliability and performance? What will be the schedule delay if I choose to implement this safety improvement? These are just a few examples.

Items such as cost, reliability, performance, schedule, and safety are called attributes of a decision when a DM uses these factors to choose among alternative courses of action. A DM evaluates each alternative course of action to determine which one best meets the objective, which varies depending on the DM's perspective. For example, a DM involved in developing a manned spacecraft would want to evaluate alternative designs based on attributes such as crew and public safety risk, cost, schedule, ability to meet science objectives, and ability to launch during the available launch window, among others. A probabilistic risk assessment (PRA) should be embedded in a decision process that involves the safety of high-consequence engineered systems.[1] This allows safety to be quantified as an attribute. Decisions that involve more than one attribute are called multiattribute decisions. A review of decision analysis literature offers a synthesis of an overarching decision process created for this book (see, for example, Keeney and Raiffa 1976; Saaty 1990; Schmid 1994; Schuyler 2001). Figure 4-1 illustrates the process, and I explain each step with examples.

4.1 Identify a Decision Opportunity

The success and failure of a project rests on the multiplicity of decisions made on almost a daily basis. In a typical engineering project, project staff members will go to one of the DMs with a problem, hoping that the DM can solve it for

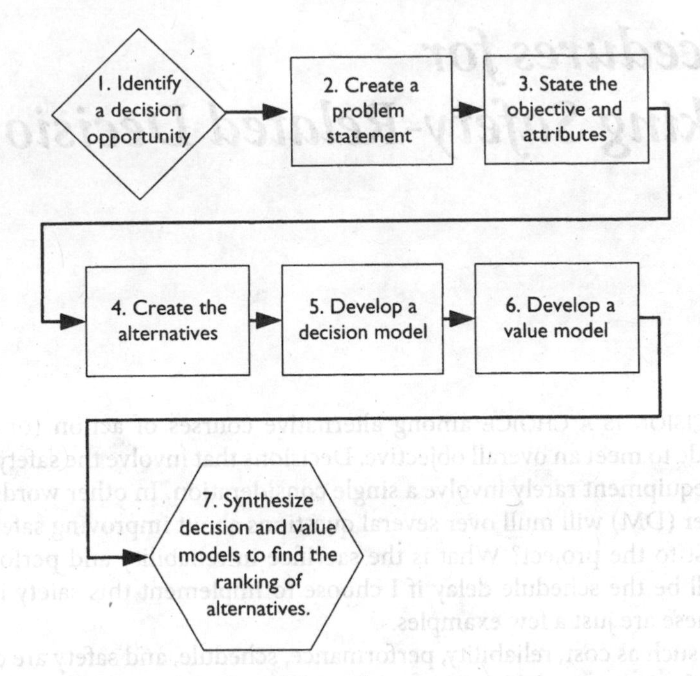

Figure 4-1. ***Decision Analysis Procedure***

them. Sometimes, the staff will present alternatives to the DM and ask the DM to choose one. Experienced DMs typically rely on intuition, past experience, and the experience of other trusted people to quickly and informally come to a decision about the problem. DMs with this kind of experience stay in lead positions because they have a track record of making reasonable decisions in this way.

Knowing when a problem becomes an opportunity is an important talent for a DM. At such times, a creative approach can have huge payoffs—fulfilling the project goals in a creative way while saving the project time and money.

Let's look at an example. Engineers were modernizing a wind tunnel by making major modifications to its design. One troublesome part of the design was a new motion control system for aircraft test models within the tunnel. The traditional method of identifying trouble spots in a design is a bottom-up approach called a failure mode and effects analysis (FMEA; DoD 1980). In this method, each component is evaluated individually and a correction or a compensating safeguard is added to the design when a failure mode is identified. In the wind tunnel example, many failure modes were identified that would allow inadvertent motion of the model, which might cause model or wind tunnel damage. The design was becoming more complicated and costly at the same time that system reliability was being compromised. Realizing that this situation could not continue, the staff brought the dilemma to the project DM.

The staff believed that burdensome safety requirements were adding cost to the project and hoped that the project DM would waive these requirements. The staff presented the DM with what amounted to an ultimatum—eliminate the safety requirements or you'll be responsible for project overruns. After a little thought, the DM recognized this as a decision opportunity. She recognized that there were at least two other alternatives: tell the staff to redesign the system from scratch or change the way in which safety analysis was done in the hope that another alternative would be revealed. She also knew that if she were to waive some safety requirements, she would have to seek approval from company directors, who would question how much risk would increase. FMEA was unable to provide this information, so the DM decided to begin a top-down PRA on the design using fault trees. A couple of weeks after the PRA began, it became apparent from the structure of the fault trees that all the failure modes were leading to the same top-level problem—that the model would impact the inner surface of the wind tunnel, damaging both. The PRA analysts pointed this out to the designers, who then quickly devised a safeguard that simply prevented the model from contacting the wind tunnel walls, no matter what the cause. This risk-reduction strategy was able to lower risk more than all the other safeguards that were previously built into the design. No safety requirements were waived, and system reliability and safety increased. Even though performing a PRA was an unbudgeted expense to the project, the design simplification resulted in net cost savings of approximately $150,000.

4.2 Create a Problem Statement

In the wind tunnel example, the staff saw the problem only in terms of safety requirements. The only options they thought of were to remove them or pay the price. But the project DM broadened the problem statement, asking "How can we make the safest design within project budget and schedule constraints?" This gave rise to new alternatives, such as looking at safety from a different perspective and changing the system design. The way in which a problem is stated often helps to create the alternatives for its solution. A powerful method for getting the most out of a decision opportunity is to examine the assumptions, constraints, and initiating mechanisms of the problem statement (Hammond et al.1999). Broader problem statements, such as those I suggest here, lead to more comprehensive alternatives. For example, an even broader way to state the problem might have been "What is the safest possible design?" Although such a statement removes the constraints posed by budget and schedule, it does not question the underlying assumption and initiating mechanism of the problem. In the wind tunnel example, the initiating mechanism was that the design was increasing in complexity and cost. Two of the underlying assumptions were that (1) the design the staff presented to the project DM was the only reasonable one and (2) safety was the only issue. A problem statement that generates even more opportunity for creative alternatives would have been something like "How can we achieve the project's objectives of producing a modern, reliable, safe, cost-effective wind tunnel motion control system?" From an even broader perspective, consider the typical strategic thought process of a company executive, who might shape the wind tunnel problem statement like this: "How can we best provide information about aircraft flight to our customers?" This type of statement calls into question the fundamental assumption that the modifications are needed or even that a wind tunnel is necessary. One sage problem solver once said that a judicious statement of the problem is 80% of the solution. Thinking creatively about the problem statement is a powerful step toward making the best safety decision.

4.3 State the Objective and Attributes

For every decision problem, a key question is "What is to be achieved by the decision?" The overall objective of the decision, which depends on the DM's perspective, reflects the answer to this question. Returning to the wind tunnel example, the business executive's objective was to produce the best and safest wind tunnel while still making a profit for the company. The project DM's objective was to ensure that the wind tunnels were safe without jeopardizing the executive's objectives. The staff's objective was to create a wind tunnel that works and meets all constraints on budget, schedule, safety, reliability, and performance.

The project DM considered two attributes to be important: reducing safety risk and staying within project budget to meet executive objectives. She chose the course of action she believed was the best overall choice—a combination of safety improvement and cost reduction. By performing a PRA, she found an alternative that reduced both cost and risk. Clearly, stating the objective and attributes is closely aligned with stating the problem. Figure 4-1 is not intended to be an immutable serial decision process. Instead, thinking about the problem, the objective, and the attributes is a holistic process that occurs naturally.

4.4 Create the Alternatives

The potential for favorable consequences of a decision is defined by the set of alternative courses of action under consideration. I call such courses of action "alternatives" in this book. DMs intuitively create alternatives when attempting to overcome a difficulty for which a decision is needed. Good decision processes, such as those described in this book, however, do not guarantee good outcomes, particularly if all creative alternatives are not actively sought.

Alternatives to reduce safety risk, which are often called risk-reduction strategies, are suggested by the event tree and fault trees of a PRA. In Figure 3-2, any design feature, operational aspect, or maintenance practice that interferes with an initiating event or impedes the ability of pivotal events to lead to unwanted end states reduces safety risk. Any feature, aspect, or practice that eliminates or reduces the severity of an end state mitigates risk. Alternative designs often emerge from thinking about the initiating and pivotal events. Because fault trees typically disaggregate a higher level of assembly (e.g., a system) into ever more detailed causal events, every level of a fault tree suggests risk-reduction alternatives. A good way to stimulate the creation of other risk-reduction strategies is to brainstorm with a diverse group of engineers and scientists who have been shown the design, the event trees, and the fault trees. The group should be diverse in terms of multiple areas of expertise and length of experience. Less experienced engineers and scientists might suggest a novel approach because they are less constrained by past experience. On the other hand, an experienced worker might remember how a similar problem was solved in the past.

During the example wind tunnel modification project, different kinds of alternatives were generated. At one level, alternatives involved changing the requirements. At another level, alternatives were generated by attempting to install safeguards for each component failure mode. The project DM, though, suggested redesign or a PRA as alternatives.

A DM often prefers to collect or generate more information before coming to a decision. When safety is involved as a decision attribute, performing a PRA generates such information. DMs should also consider the "null alternative" (i.e., doing nothing). The null alternative provides the baseline risk against which other alternatives can be compared. The gather-more-information alternative often turns out to be a good first step. Additional information can suggest more alternatives, find that some alternatives are not practical, and reduce uncertainties when scenarios and consequences are uncertain. Of course, gathering more information is worthwhile only if the cost to do this is less than the increase in value of the expected outcome. For example, if the PRA of the wind tunnel had cost more than the realized cost savings (i.e., $150,000) or more than the expected cost of potential damage to the models or the wind tunnel, it would not have been worth the effort.

As I already mentioned, each rephrasing of a problem statement amounts to a reexamination of constraints and assumptions. Consciously identifying implicit and explicit constraints and assumptions, then, is fruitful, allowing more alternatives to come to mind.

During the wind tunnel project, the executives thought of alternatives, such as developing more powerful simulation or computational models, renting time on other wind tunnels to reduce the high infrastructure costs, modifying existing wind tunnels, scrapping wind tunnels altogether, or building new wind tunnels. These are quite different from the alternatives considered by the project DM and design staff.

4.5 Develop a Decision Model

As I described in Chapter 3, a PRA shows how a system responds to perturbations. It does this in the form of scenarios that describe the various potential sets of events that might occur, the causes of these events, and the consequences resulting from each set of events. PRA attempts to find all significant future scenarios, but the analyst does not know which events will take place because knowledge of the future is uncertain. This uncertainty is described by showing (see Figure 3-4 for an example) alternative scenarios, each characterized by a probability of occurrence.

4.5.1 The Role of the Decision Model

A decision model attempts to grapple with an uncertain future as well. It attempts to estimate the consequence or outcomes resulting from each hypothetical alternative course of action. Such a model shows the potential responses to selected alternatives. Each alternative in a decision, just like each perturbation in a risk model, can have more than one outcome because of events that might or might not occur in the future after the decision is made. This is illustrated in the decision tree shown in Figure 4-2.

The round node in Figure 4-2 at the split between the possible occurrences of more than one event is called a chance node. In the figure, the chance node introduces two events: "Leak and No leak." A chance node in a decision tree is similar to a pivotal event in an event tree. Both depict alternative events and alternative paths governed by chance. The events extending from a chance node or pivotal event must be mutually exclusive and exhaustive for a valid analysis. This decision tree also has decision nodes, shown as square nodes, that represent the point in time at which a decision is to be made. Following a decision node are alternative courses of action such as Test or No test. A decision tree ends up on the right-hand side with the probability of each path and the consequences of taking that path.

4.5.2 Initial Field Joint Decision Model

In this section, I explain the development of a decision model by presenting an example about the development of field joints of a new solid rocket motor (SRM) for a launch vehicle.

Problem description. Engineers were designing and developing a new SRM to provide more thrust for heavier payloads for a manned launch vehicle. As with many large SRMs, this was a segmented design, in which segments attach to each other using what is called a "field joint." The SRM would require assembly at the launch pad. The project staff considered three alternative field joints: the same tang and clevis design as the old SRM (the "old" design pictured in Figure 4-3), an improved tang and clevis design, and an entirely new concept called a "flange joint," which is pictured in Figure 4-4. In addition to improving performance, improving the safety of the SRM was a key objective. A critical safety question was whether the propulsive hot gas could leak through the field joint. If this were

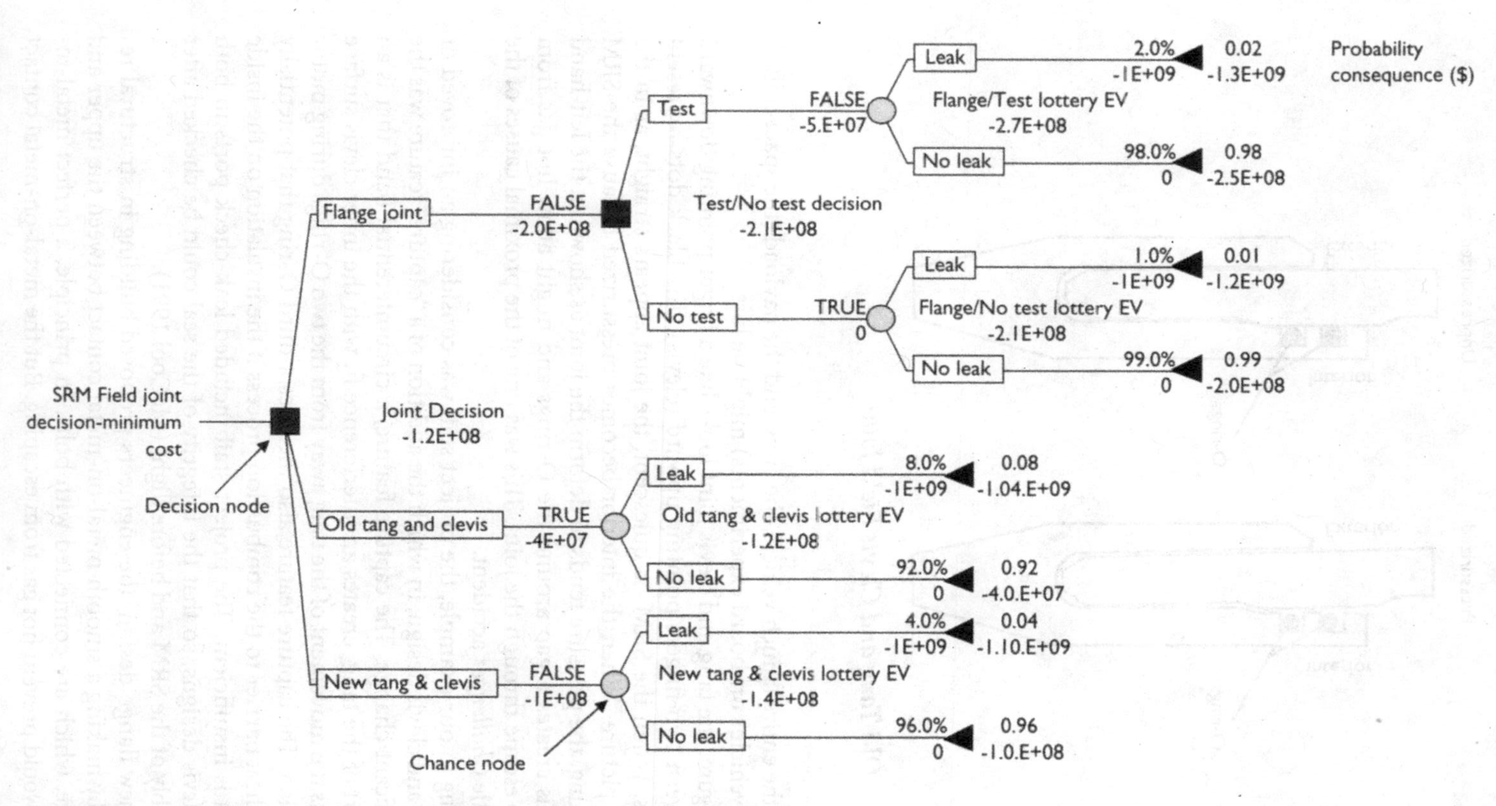

Figure 4-2. *Decision Tree for Solid Rocket Motor (SRM) Field Joint Decision*

Notes: EV = expected value. Decision rule is to minimize cost.

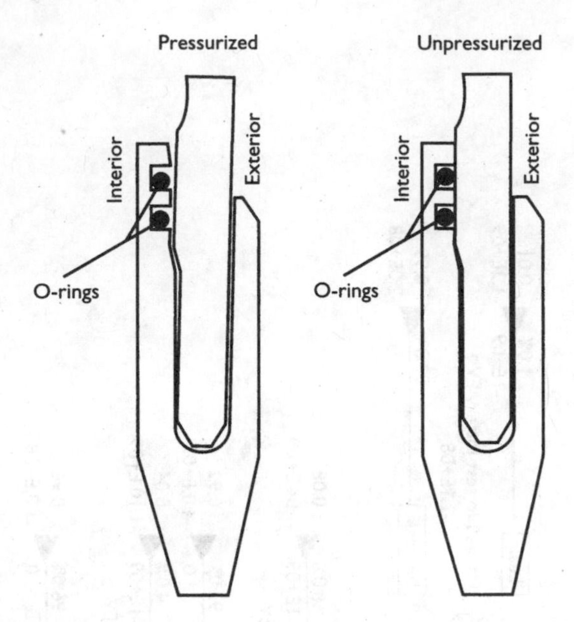

Figure 4-3. ***Old Tang and Clevis Field Joint***

to happen, the entire launch vehicle, the crew, and the payload (the spacecraft or the launch vehicle's on-board experiments) might be lost.

In the figure, the tang and clevis joint looks like a finger pointing downward (tang) between two fingers pointing upward (clevis). The black dots represent the O-rings. When the SRM is quiescent, the joint appears straight, as in the right-hand picture. When the interior becomes pressurized because the SRM fuel is burning, the pressure tends to deform the joint as shown in the left-hand picture. This creates a gap around the O-rings and might allow hot gas from the SRM to escape through the joint. This was one of the proximal causes of the Space Shuttle *Challenger* accident.

Returning to our example, the project staff was considering an improved or "new" tang and clevis design, in which the addition of a "capture" feature was the most significant change. The capture feature, a circumferential band that is an integral part of the tang, creates an interference fit with the inner clevis surface and restricts the movement of the tang away from the two O-rings during motor pressurization. The capture feature also includes a third O-ring that potentially serves as a heat barrier to the combustion process if the insulation on the inside of the joint is insufficient. The project staff included leak-check ports in both tang and clevis designs, so that the integrity of the seal could be checked after field assembly of the SRM and before flight (McCool 1991).

In the new flange design, the engineers proposed building in structural redundancy by making a smooth metal-on-metal contact between the upper and lower flange, which are connected with bolts. In principle, a perfect metal-to-metal seal would prevent hot gas from escaping. But the metal-on-metal contact

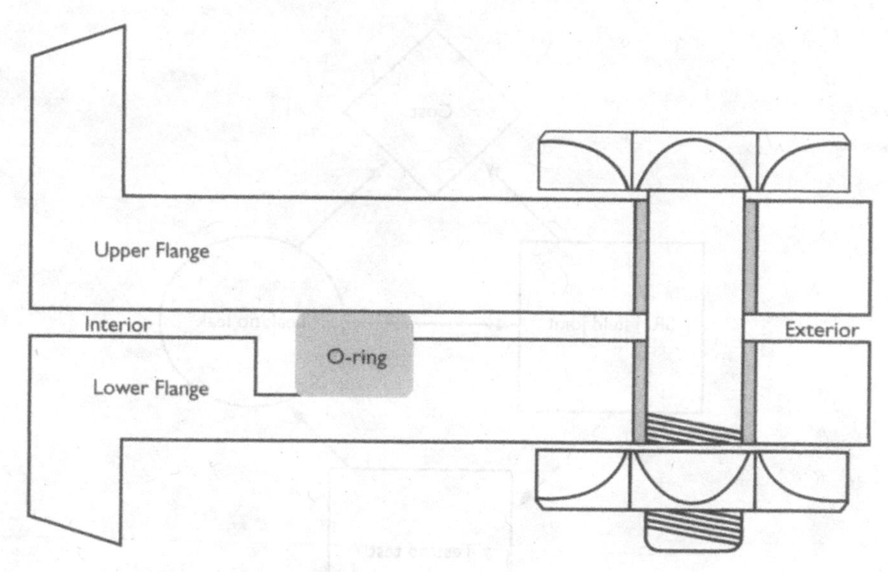

Figure 4-4. *Flange Field Joint Design*

would be subject to contaminants and differential deformation that would reduce the joint's effectiveness as a seal. In our example, then, the project team designed O-rings as a backup method to stop such leaks. Only one of three proposed O-rings are pictured in Figure 4-4.

Some project engineers argued that because a flange joint was inherently a tight metal-to-metal seal, it did not require preflight tests to check for leakage. Others argued that one could not be sure that the joint segments were mated correctly during field assembly, and that because an error in mating the segments would be sufficient to cause leakage, a preflight leak check would be prudent. Figure 4-4 does not picture the leak-check ports that would be required to test the seal before flight. These ports are essentially capped tubes that are bored through the flange metal from the outside of the joint into the spaces between O-rings. The ports add cost to the manufacture of the joint and provide additional leak paths for hot gas.

Influence diagram. The designers asked the project DM to decide what to do about the flange joint. Should the joint be tested for leaks or not? The project DM gave the matter some thought and realized that he should really be addressing a broader issue: "Which SRM field joint should be chosen?" He hired a decision analyst to help him to address the design team's question about testing the flange joint design, as well as his own broader concerns. Together they constructed the simple influence diagram shown in Figure 4-5.

The diagram succinctly shows the key factors of a decision and the direction of the arrowed lines (called "arcs") shows the direction of the influence (Shachter 1988). The diamond represents the consequence of a decision. In this case, the project DM wanted the consequences depicted as costs in dollars. The decision

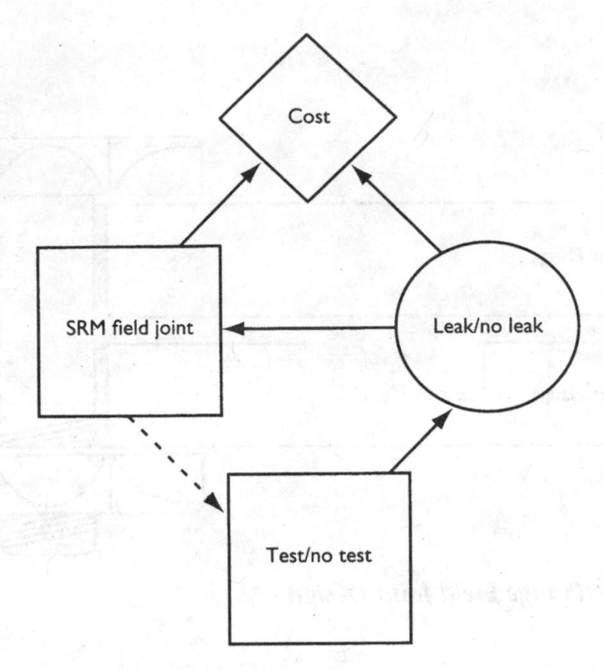

Figure 4-5. ***Influence Diagram for Field Joint Decision***

Note: SRM = solid rocket motor.

node (square) labeled SRM field joint depicts the broad question of which joint design to pursue. The analyst and the project DM reasoned that, to evaluate the flange joint, the DM should also explicitly consider the hypothetical decision about whether to leak test or not (depicted by the square labeled Test/no test.) The broken arc in Figure 4-5, between the two decision nodes, depicts an asymmetric influence. If the DM were to select the flange joint, he would need to make a second decision about the leak testing. Selecting either tang and clevis option would not require such a decision because tang and clevis designs are always tested. The chance node (circle) depicts the uncertainty about hot gas leaking. For the DM, this reflected the key safety issue—leaks might or might not occur for any joint design. Whether the joint is tested or not influences the probability of leak. In this example, the probability and consequences of a leak influenced the selection of the joint. The joint selection and the probability of leakage and its consequences influenced the cost of each alternative. Through this process, the original question brought by the design team became embedded within a larger decision model that addressed the needs of the project rather than the needs of the design team.

Decision tree. The decision analyst used the influence diagram of Figure 4-5 as a quick construct to depict the influences among nodes in a succinct, visual way. He understood that the difficulty with such a diagram is that the analyst cannot visualize the consequences associated with each path because the paths are not individually shown. Because an influence diagram and a decision tree constitute a homomorphism (in that an equivalent decision tree exists for each influence diagram), an equivalent decision tree was easily developed. Figure 4-2 shows all the paths of concern to the decision stemming from the square decision node at the left of the diagram.

Each tang and clevis alternative is followed by a chance node (circle) with events Leak or No leak. The probabilities of leaks for each alternative are shown above the horizontal lines (i.e., the branches extending from a chance node). In looking at the decision tree, the decision analyst realized that he needed help in estimating the leak probabilities. He asked the project DM who, in turn, called for a PRA to be performed using a combination of historical records and engineering analysis. The resulting PRA indicated that there would be an 8% chance of leakage for each launch using the old tang and clevis joint. The PRA also resulted in estimates for the improved tang and clevis joint, as well as for the proposed flange joint. The PRA revealed that both would be expected to reduce the probability of leak—to 4% for the new tang and clevis joint and to 2% and 1% for the tested and untested flange joint, respectively.

The flange joint alternative in Figure 4-2 is followed by the second decision node, which seeks to evaluate whether a leak test before launch would be required. Each of the alternatives—test and no test—is also followed by the chance events that represent the uncertainty of leak after launch. The analysis found that a flange joint, which was tested before launch, had a 2% chance of leaking hot gas during flight. In a surprising result, the PRA revealed that not testing the flange joint would be safer (having a lower leak probability) than testing the flange joint. The flange joint can be designed so that the segments mate properly in only one way, making it obvious when the segments are not mated. Testing the joint during or after the mating process, then, has little benefit, and the testing procedure itself introduces the possibility of contaminating the joint and the potential for leakage is sensitive to debris left in the joint. In addition, the test ports add additional hot gas leak paths. Using the PRA as a guide, the analyst and the project DM decided that the disadvantage of testing would overcome the advantage in this case.

Each branch (i.e., horizontal line) extending from a chance node requires two inputs: a probability of occurrence and an attribute amount. In our example, the project DM decided to analyze this decision tree using the cost of each alternative as the attribute of interest. The maximum cost of a leak is the loss of the vehicle, the crew, and the payload. Initially ignoring the cost of lives saved, he assigned a cost of one billion dollars to replace the vehicle and payload. This is depicted in Figure 4-2 as −1E+09 below the leak branch, where a negative number indicates cost. There is no additional cost if a leak does not occur.

Each branch extending from a decision node requires a cost. In Figure 4-2, this is shown as a number below the horizontal line (or branch) extending from the decision node. In this case, the cost of developing the flange joint was approximately $200 million, the cost of developing a new tang and clevis joint was approximately $100 million, and the cost of adapting the old tang and clevis design to the new SRM was approximately $40 million. The analyst set the additional cost of developing a testable flange joint and the cost of testing before flights at approximately $50 million.

End nodes in this decision tree are represented by triangles pointing to the left. An end node has a probability and a consequence, and a probability is associated with each chance node in a path. The consequence of a path (cost, in our example) is the sum of the costs in each path. For example, the total cost of the old tang and clevis leak path is 1.04E+09, which is the sum of 1E+09 and 4E+07 (shown in Figure 4-2 as –1.04E+09 below the 0.08). The probability entry generally represents the product of the chance nodes in a path. In this path, there is only one chance node. Therefore, the probability of this path is 0.08.

A decision tree is evaluated from right to left by successively collapsing the tree. Each chance node presents what decision analysts call a "lottery." A lottery has an expected value (EV) given by the sum of the products of the probabilities and consequences of each branch as follows:

$$EV = \sum_{N} P^{i} C^{i} \tag{4-1}$$

where there are $i = 1, \ldots, N$ branches following a chance node in a lottery. The EV of the lottery replaces the chance node on the decision tree. For example, the EV of the flange/test lottery in Figure 4-2 equals (0.02 × –1.3E+09) + (0.98 × –2.5E+08) = –2.7E+08.

Figure 4-2 shows the lottery EVs for our example in between the horizontal lines following a chance node. The decision analyst collapsed the tree by substituting the EVs of the lotteries for the chance nodes, yielding the tree shown in Figure 4-6, which is mathematically equivalent to the decision tree in Figure 4-2.

At first, the project DM had decided on a decision rule—to minimize cost. For the test/no test decision, the least-cost path was to skip testing of the flange joint, a path shown with a TRUE in Figures 4-2 and 4-6. Consequently, if the project DM simply needed to provide guidance to the flange designers, this analysis would indicate that the better course of action would be not to leak test the flange. But the project DM understood that the decision was broader than that. Using the same decision rule to minimize cost, the decision would be to adapt the old tang and clevis joint to the new launch vehicle, even though the probability of losing the launch vehicle and payload is the highest of all alternatives. This path is also shown by a TRUE in Figure 4-6, and the expected value of the decision is shown under joint decision. The EV of this decision—which is a cost—would be $120 million. The flange joint/test alternative would cost $270 million, the flange joint/no test alternative would cost $210 million, and the new tang and clevis alternative would cost $140 million.

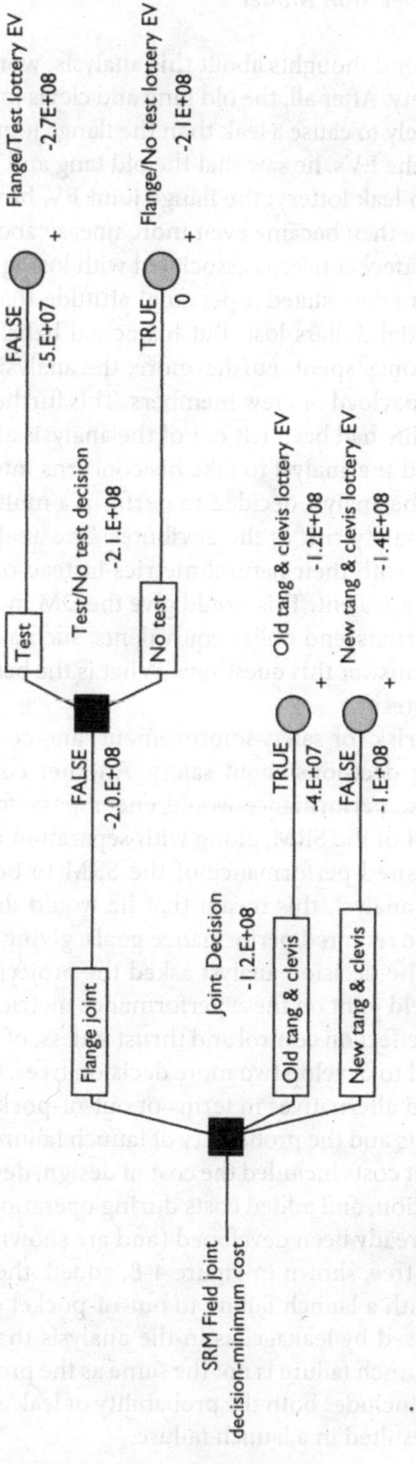

Figure 4-6. ***Collapsed Decision Tree after Evaluating Lotteries***

Note: Decision rule is to minimize cost.

4.5.3 *Multiattribute Decision Model*

The project DM had second thoughts about this analysis, wondering if it placed enough emphasis on safety. After all, the old tang and clevis joint is between four and eight times more likely to cause a leak than the flange joint. Looking further into the contributors to the EVs, he saw that the old tang and clevis EV is driven by the EV of the leak/no leak lottery; the flange joint EV, however, is driven by the development cost. He then became even more uneasy about the analysis because, by including the safety concerns associated with losing the launch vehicle as a cost only, he had in effect stated a personal attitude that equates safety to dollars spent and potential dollars lost. But his actual belief was that safety is more important than money spent. Furthermore, the analysis did not consider the impact of losing the payload or crew members. This further added to his discomfort because loss of life had been left out of the analysis altogether.

The project DM asked the analyst to take his concerns into account in a second decision analysis. The analyst decided to perform a multiattribute decision analysis using cost and safety risk as the attributes. The analyst would attempt to portray the attributes with their natural metrics instead of converting all attributes to a monetary equivalent. This would give the DM more flexibility to account for concerns that transcend dollar equivalents, such as possible fatalities. The DM was looking to answer this question: "What is the best alternative when considering both attributes?"

The attributes safety risk (or safety improvement) and cost are the most frequently used in making decisions about safety. Another common attribute is performance. In this case performance would encompass, for example, thrust, acceleration, and control of the SRM, along with separation at the proper time. The DM wanted the desired performance of the SRM to be treated as a constraint. To the decision analyst, this meant that he would discard any alternative that did not meet the required performance goals, giving that alternative no further consideration. The decision analyst asked the project engineering team about the effect of the field joint on these performance metrics and was told that the joint would have no effect on control and thrust unless, of course, there was a leak. The analyst decided to develop two more decision trees. One tree, shown in Figure 4-7, characterized alternatives in terms of out-of-pocket costs associated with developing the joints and the probability of launch failure, p(LF), as the two attributes. Out-of-pocket costs included the cost of design, development, testing, evaluation, implementation, and added costs during operation of the launch vehicle. These costs had already been developed (and are shown in Figure 4-2).

The second decision tree, shown in Figure 4-8, added the expected replacement costs associated with a launch failure to out-of-pocket costs. Expected replacement costs are caused by leakage, as in the analysis that produced Figure 4-2. The probability of launch failure is not the same as the probability of leak occurrence because p(LF) includes both the probability of leak and the conditional probability that a leak resulted in a launch failure.

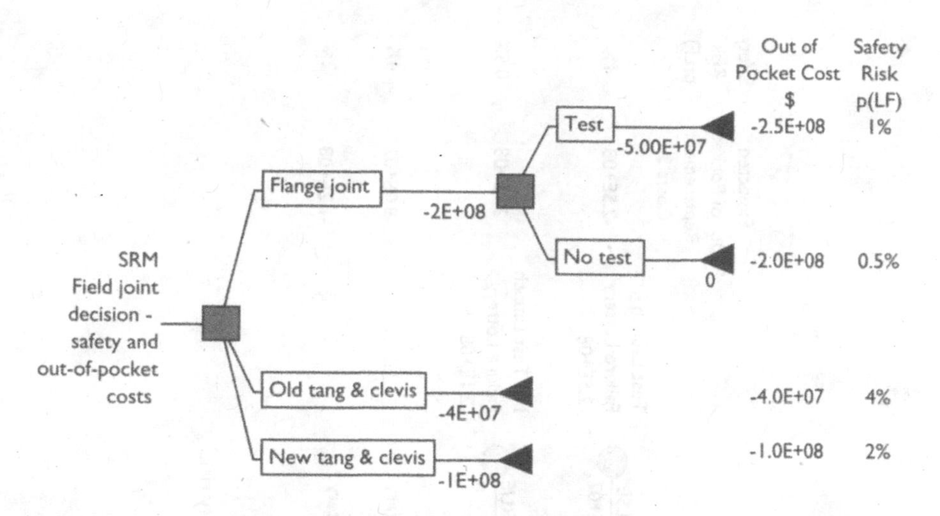

Figure 4-7. ***Second Decision Tree for Solid Rocket Motor (SRM) Field Joint***

Note: Decision rule is to select the best combination of out-of-pocket costs and safety risk.

In Figure 4-7, the consequences of each path, C^i, in the tree were represented as a vector, $[X^i]$, of n attribute values, y^j, where Y^j represents attributes. In this example, the Y^j are out-of-pocket costs represented in dollars and safety risk represented by p(LF). In this example, the analyst represented the consequences of multiattribute decisions as follows:

$$C^i = [X^i] = [y_1,\ldots,y_n]^i \quad (4\text{-}2)$$

For example, the vector $[X^i]$ for the path flange joint/test was equal to [–$2.5E+08,1%].

4.6 Develop a Value Model

"A value model v assigns a number $v(x)$ to each consequence $x = (x_1, \ldots, x_N)$, where x_i is a level of attribute X_i measuring objective O_i, such that the numbers assigned both (1) indicate the relative desirability of the consequences and (2) can be used to derive preferences for alternatives" (Keeney 1992, *129*).

A value model represents a DM's preferences associated with the magnitudes of the attributes. For example, is half as much safety half as desirable? Is twice the cost twice as objectionable? A value model also represents the DM's attitudes toward the attributes. For example, how much more or less does the DM value safety relative to cost? How much safety will a DM give up to reduce cost? In the example I give in this section, the attributes of interest will continue to be cost and safety risk.

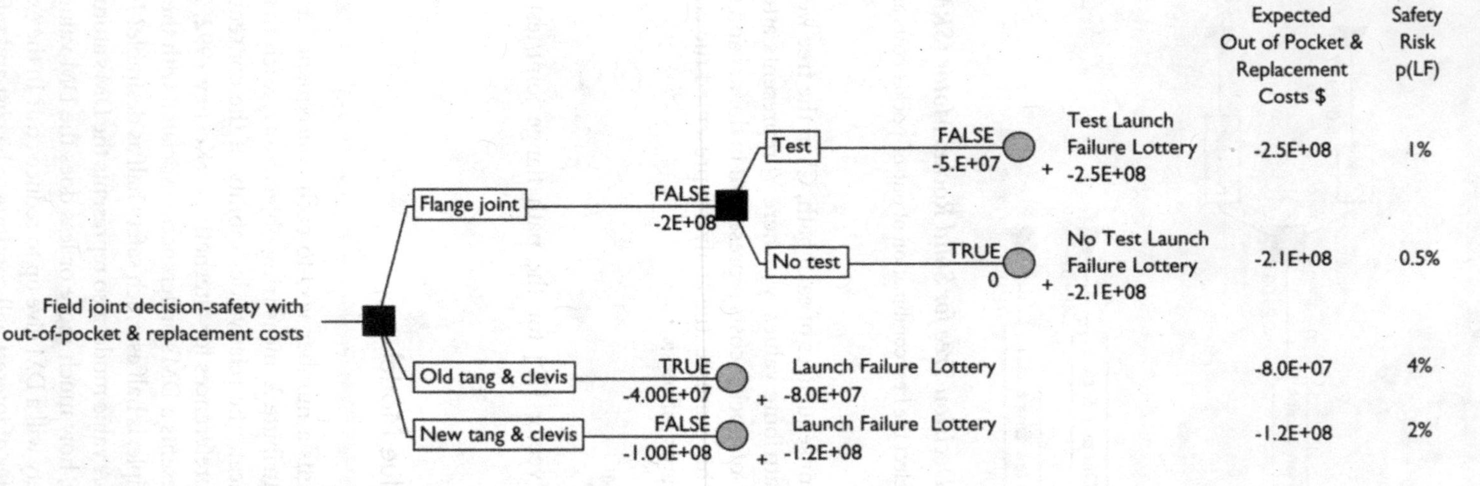

Figure 4-8. ***Solid Rocket Motor (SRM) Field Joint Decision Tree***

Note: Decision rule is to select the best combination of out-of-pocket and replacement costs and safety risk.

On Uncertainty and DM's Preferences

For pictorial clarity, I presented the consequences and probabilities in this section as single values, sometimes called point estimates (the term *point estimate* pertains to a single number that represents an uncertain quantity). Clearly, however, the attribute values of cost and safety risk exhibited epistemic uncertainty (discussed in Chapter 2) because they represented the PRA analyst's best estimate of potential future events. For example, the uncertainty in the safety risk numbers of Figure 4-8 were developed during the PRA and are represented by the set of curves in Figure 4-9.

The vertical lines in Figure 4-9 illustrate the location of the mean values shown in Figure 4-8. Note that the curves overlap and the new designs have larger uncertainty than the old tang and clevis design because at this point in the decision process, there is no operating experience with the new designs. Although the mean values indicate that the new designs will be safer, there is no guarantee that it will turn out that way in operation. The overlaps, therefore, indicate the uncertainty about which option is safer, that is, which option has the lower probability of launch failure.

Figures 4-7 and 4-8 are example implementations of the decision concept shown in Figure 1-1. Ultimately, the DM must decide how he values safety risk against out-of-pocket expenses. In order to make a cogent decision that reflects his values, the DM must also decide on the relative merits of different levels of cost and safety. As discussed in detail in the next sections, a typical DM may have strong opinions about cost

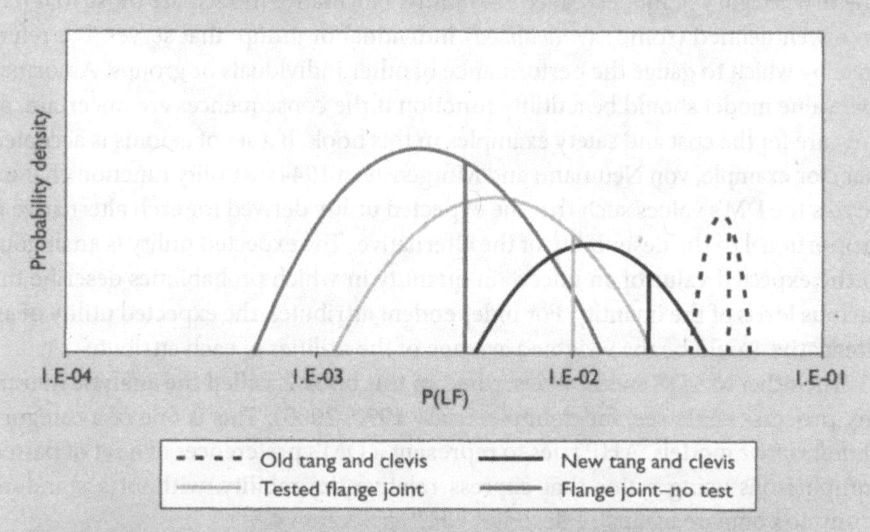

Figure 4-9. ***Probability Distributions for Probability of Launch Failure, p(LF)***

increases and safety improvements. For example, a DM may believe that a cost of $200 million is much worse than a cost of $150 million, not just 33% worse, because $200 million exceeds his budget. On the other hand, the DM may be relatively insensitive to costs below $100 million because such costs may be within his budget. With respect to safety, a similar nonlinearity in the value of the probability of launch failure would be typical. A DM may be quite sensitive to launch vehicle failures that approach 5% because the industry average launch vehicle failure rate is approximately 5%. If these undesirable limits are taken as constraints, the old tang and clevis alternative and both flange joint alternatives would be eliminated, leaving only the new tang and clevis alternative.

The intent of the formulations of Figures 4-7 and 4-8 is to keep safety issues apart from cost issues in order to allow the project DM to reflect his view of the relative importance of cost and safety.

To develop a value model, the analyst must elicit these attitudes from the DM and represent them mathematically. The value model substitutes for the consequences in Equation 4-2 and the decision tree. In other words, $C^i = [X^i] = [y_1, \ldots, y_n]^i$ becomes $V(C^i) = V[X^i] = [V(y_1),\ldots,V(y_n)]^i$, where V stands for "value of."

I describe two commonly used multiattribute decision analysis (MADA) methods in this book, each with its own value model. One, called multiattribute utility theory (MAUT) is based on utility theory (see, for example, Keeney 1976), which is one of a category of models called *normative*. Normative models are those that rely on a well-defined (some say *idealized*) individual or group that serves as a reference by which to gauge the performance of other individuals or groups. A normative value model should be a utility function if the consequences are uncertain, as they are for the cost and safety examples in this book. If a set of axioms is accepted (see, for example, von Neumann and Morgenstern 1944), a utility function characterizes the DM's values such that the expected utility derived for each alternative is proportional to the desirability of the alternative. The expected utility is analogous to the expected value of an uncertain quantity in which probabilities describe the various levels of the quantity. For independent attributes, the expected utility of an alternative would be the weighted average of the utilities of each attribute.

The other MADA method described in this book is called the analytic hierarchy process (AHP; see, for example, Saaty 1990, 2000). This is one of a category of *descriptive* models. AHP tries to represent a DM's preferences as a set of paired comparisons using ratios that express relative desirability, without a standard group to compare against. I describe AHP in Section 4.8.

Using either model in a multiattribute decision problem, however, requires additional assumptions. For example, analysts typically assume that preferences are independent. If the preferences are independent, a DM's attitudes about the

Risk "Personalities"

A utility function is also a value model but additionally represents the DM's attitude toward risk. In other words, a utility function represents the DM's preferences and the degree to which she is risk averse, risk neutral, or risk prone. Suppose there are two alternatives. One yields a sure $1 million. The other alternative offers a 50% chance to achieve $2 million and a 50% chance of gaining nothing. The expected value of each alternative is $1 million. Decision analysts call someone who is indifferent about the two alternatives a "risk-neutral" individual. A risk-neutral person's attitude toward the 50/50 lottery is the same as that toward the sure yield. A risk-averse person would value (or desire or prefer) the sure alternative to the lottery. A risk-prone person (i.e., a gambler or risk taker) would take the chance of achieving the higher payoff and accept the fact that she might get nothing. In other words, she would prefer the lottery (or the gamble) rather than the sure thing.

safety of a risk-reduction alternative such as an improved field joint would not be influenced by his knowledge of the field joint's cost.[2]

4.6.1 *Utility Function Concepts*

The common way to denote a utility function is u(y), where y is an attribute consequence level (or more than one). For example, a DM's preferences about different safety risks and costs—which are derived from his set of principles or values—can be represented by a function over safety risk levels and a function over costs. According to utility theory, the DM will prefer consequence y_1 to consequence y_2 if $u(y_1) > u(y_2)$.

In Figures 4-10 and 4-11, I show example shapes of utility functions for risk-averse, risk-neutral, and risk-prone individuals. Analysts typically normalize utility numbers such that they are between 0 and 1, although this is not absolutely necessary. Figure 4-10 depicts preferences (or utility numbers) that increase with increasing attribute magnitudes—such as revenue, safety, profits, or winnings—where more is better. Risk-averse utility curves for increasing functions are always concave (Keeney 1992), which means that the function has a decreasing slope as the attribute magnitude increases.

The shapes of the curves make intuitive sense. The risk taker in Figure 4-10 assigns relatively little value to the lower end of the revenue scale, but the utility increases rapidly as revenues increase. In contrast, the risk-averse DM is satisfied with lower revenue increases and reluctant to gamble to obtain large revenues. The risk-neutral DM assigns equal value to any increment of revenue.

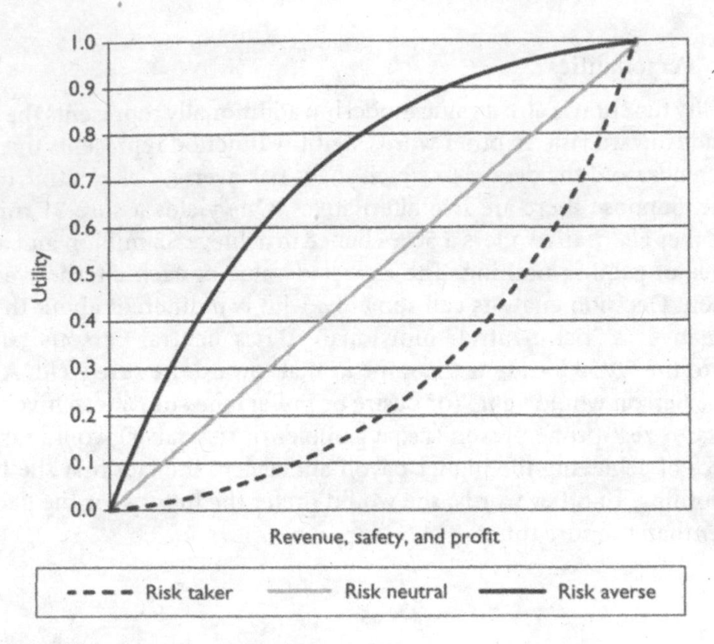

Figure 4-10. *Increasing Utility Functions*

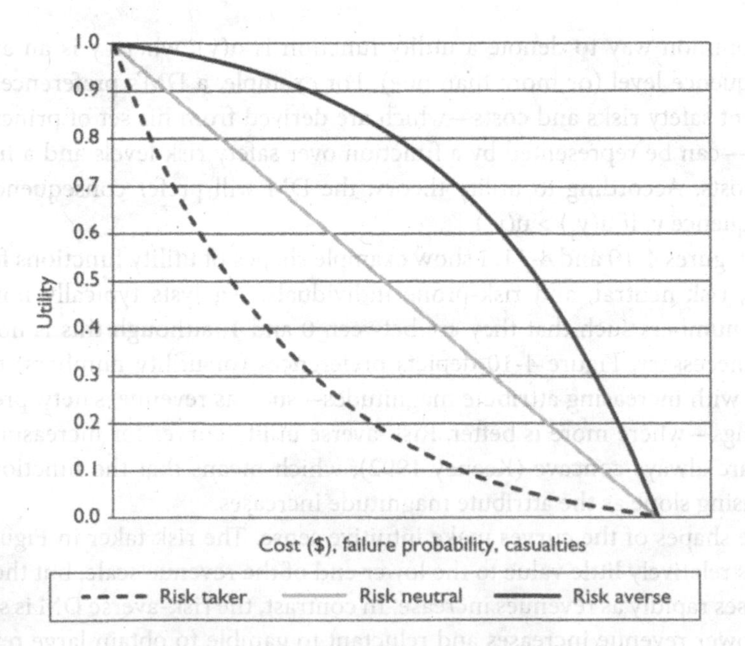

Figure 4-11. *Decreasing Utility Functions*

We can think about these three types of DMs in another way—through the concept of *certainty equivalence.* As we've established, there's usually uncertainty in making decisions. The certainty equivalent (CE) of a decision in which there is a possible gain or loss is the attribute amount (e.g., monetary amount) that the DM is willing to settle for to eliminate the uncertainty. For example, in a 50/50 lottery in which a gain of $X or a loss of $Y can occur, the CE is the amount of money for which the DM is willing to *sell* the lottery. In other words, at this amount the DM is indifferent between the lottery and the certain gain. For increasing attribute magnitudes, such as in Figure 4-10, (1) a risk-averse DM wants to avoid the gamble and is willing to sell the lottery for less than its expected value (i.e., a risk-averse CE is less than ($X+$Y)/2); (2) a risk-neutral DM would sell the lottery for its expected value; and (3) a risk-taking DM would withhold selling unless he would receive more than the expected value. The third attitude indicates that the DM is willing to gamble to potentially receive a larger gain.

Figure 4-11 illustrates the three types of DMs for decreasing utility functions such as probability of failure, number of casualties, and monetary loss. For these utility functions, the CEs for the three types of DMs are reversed. The risk-averse DM will *pay* more than the expected value to avoid a potentially larger loss. The risk taker is optimistic about his chances of having a smaller loss and will pay less than the expected value.

Figure 4-12 shows an example of attitudes toward safety as characterized by probability of launch success, p(S). This DM's preferences vary from risk prone at low probability of launch success (low safety) to risk averse at high probability of launch success (high safety). For a project DM developing a launch vehicle, this set of preferences is understandable. When the probability of success is low, a DM is willing to gamble to increase it. The lower left side of the curve, therefore, reflects a risk-taker utility function. As the probability of launch success rises to

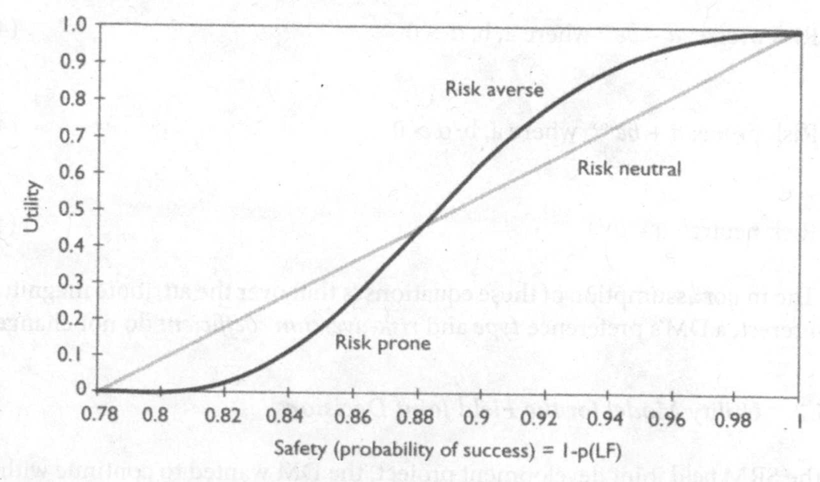

Figure 4-12. ***Risk-Prone to Risk-Averse Utility Function***

levels that are considered "safe enough," the DM is less willing to gamble to increase it. This is particularly true in light of the uncertainty in the ultimate level of safety (as indicated by Figure 4-9). The right side of the curve in Figure 4-12, therefore, reflects risk aversion.

To obtain utility curves, the decision analyst typically interviews a DM in a manner similar to the knowledge elicitation I covered in Section 3.6.2. To probe a DM's risk preferences, the interviewer asks one or more questions to establish the ranges of the attribute magnitudes (e.g., p(LF), project cost) for which a DM is risk prone, risk averse, or risk neutral. Many decision analysts have found that using an exponential for risk-averse and risk-prone attitudes results in sufficient accuracy for many decisions. Of course, a risk-neutral function is always represented by a straight line on a linear graph.

The following functions of the attribute variable, x, are typical of increasing utility functions (Kirkwood 1991):

Risk averse: $a - be^{-\alpha x}$ where $a, b, \alpha > 0$ (4-3)

Risk prone: $a + be^{\alpha x}$ where $a, b, \alpha > 0$ (4-4)

Risk neutral = $a + bx$ (4-5)

In this formulation, α represents a scale factor that is typically called a risk-aversion coefficient. Its reciprocal, ρ, is called risk tolerance. The coefficients, a and b, are adjusted to scale the utility function to have magnitudes between 0 and 1 over the domain of interest. Typical equations for decreasing utility functions follow:

Risk averse: $a - be^{\alpha x}$ where a, b, $\alpha > 0$ (4-6)

Risk prone: $a + be^{-\alpha x}$ where a, b, $\alpha > 0$ (4-7)

Risk neutral: $a - bx$ (4-8)

The major assumption of these equations is that over the attribute magnitudes of interest, a DM's preference *type* and *risk-aversion coefficient* do not change.

4.6.2 Utility Model for the Field Joint Decision

In the SRM field joint development project, the DM wanted to continue with the analysis represented in Figure 4-8 because it was the most comprehensive of the three developed by the decision and PRA analysts. It also satisfied his perspective that safety should be presented separately from costs. To establish the preference type and the risk tolerance, the decision analyst interviewed the DM.

During the interview, the project DM decided that any new design should not be less safe than the old tang and clevis design. In other words, the probability of success of the vehicle resulting from field joint failures, p(S), should not be any lower than 96%. In fact, he would have liked to make this a constraint, but in looking at the uncertainties in Figure 4-9, he recognized that this would not be feasible.

The decision analyst asked two key questions, one about safety and one about cost. First, the analyst asked: "Given a lottery in which there is a 50/50 chance of having an *s*% gain in p(S) or a 2% loss in p(S), what is the magnitude of *s* for which you are indifferent?" The expected value of this lottery is 0.25*s*. The indifference amount of *s* has a CE of 0. At this amount, $s \cong \rho = 1/\alpha$ (Kirkwood 2002).

The DM studied the uncertainties in Figure 4-9 and saw that the old tang and clevis design could turn out to be as bad as 95% (an approximate 1% loss) and as good as 97% (an approximate 1% gain). Because his criterion was to do better than this design, his indifference point for this hypothetical lottery was equivalent to $s = 2\%$. He was comfortable with a design with an expected gain of 0.5%. The decision analyst learned two lessons from this answer. First, the project DM is mildly risk averse with respect to this attribute. Second, the amount of $\alpha = 0.5\%$ in Equation 4-3. After normalizing to obtain a and b in Equation 4-3, the DM's safety utility function became apparent and is shown in Figure 4-13.

A similar procedure was used to determine the DM's cost preferences. In this case, the DM found it easier to directly provide the two end points on the utility curve. He said that out-of-pocket or total expected costs greater than $200

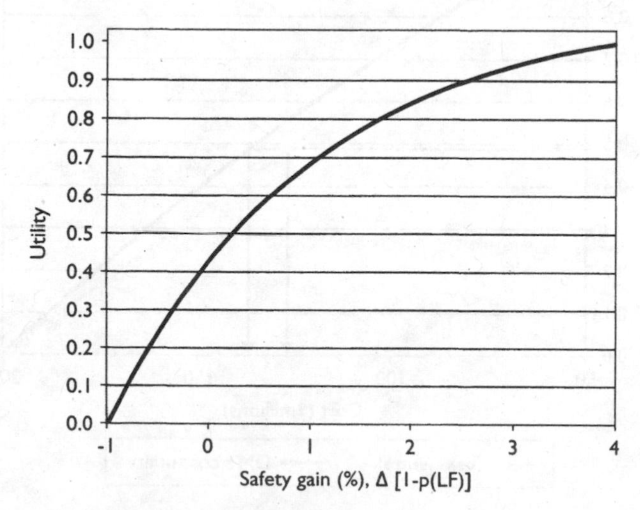

Figure 4-13. ***Field Joint Safety Utility Function***

million are extremely undesirable. The most desirable costs would be under $50 million, but he would be satisfied if costs stayed under $150 million. To clarify the last statement, the decision analyst asked the DM, "How much of a hypothetical insurance premium would you be willing to pay to avoid a gamble over a 50% chance of a $200 million cost and a 50% chance of a $50 million cost?"

The DM mulled this over and, being risk averse with respect to gambling with costs, he said that $135 million would be reasonable. This dialogue yielded three points that correspond with three unknowns in Equation 4-6. Two points are

Substituting Utility for Consequences

In the first part of Section 4.6, I describe two types of value judgments a DM must make to develop a normative decision model. The first is associated with development of utility functions which substitute for consequences when the DM wants to include her values in the decision. If values are described in terms of utilities, then, for each consequence level, i:

$$C_i = [X_i] = [y_1, \ldots, y_n]' \text{ becomes}$$
$$U(C_i) = U[X_i] = [u(y_1), \ldots, u(y_n)]' \quad (4\text{-}9)$$

The second value judgment relates the DM's relative importance of one attribute to another. I describe this in Section 4.7.

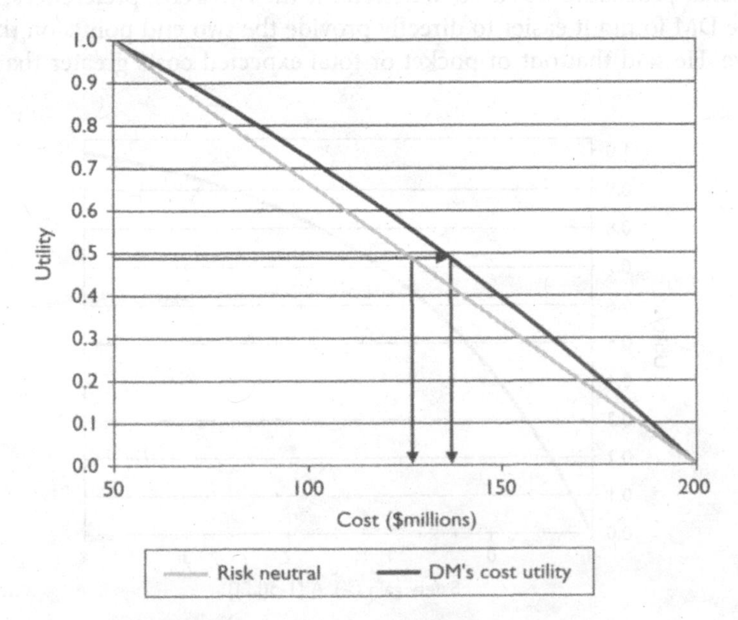

Figure 4-14. *Field Joint Cost Utility Function*

0 utility at $200 million and a utility of 1 at $50 million. Figure 4-14 illustrates the third point, along with the project DM's cost utility function. The expected value of the lottery under scrutiny is $125 million, which would correspond to a utility of 0.5 for a risk-neutral DM because it is equidistant between the upper and lower bound. This DM said, though, that he would be willing to spend $135 million to avoid the lottery, revealing his indifference level. This is a risk-averse position. Because the DM is indifferent to $135 million or the lottery, his utility for $135 million must be the same as that of the utility of the EV of the lottery of the risk-neutral DM (von Neumann and Morgenstern 1944).

Table 4-1 shows the consequences and their associated utilities for the SRM field joint decision tree in Figure 4-8. These utility numbers were derived from Figures 4-13 and 4-14.

The decision analyst substituted expected utilities for the calculated cost and safety risk in Figure 4-8, yielding Figure 4-15.

4.7 Synthesize Models and Rank Alternatives

How can the expected utilities in Figure 4-15's decision tree be combined to arrive at an overall decision about which alternative is best? To do this, we must know the relative importance of cost and safety to a DM, who must provide this information within the context of this decision. We don't need to know the DM's overall philosophical opinion about the relative importance of cost and safety in the world. Only his attitudes about cost and safety *relative to this decision* are relevant.

Much theoretical discussion centers on how to combine utilities of multiple attributes into a cogent decision methodology in a mathematically rigorous way (see, for example, Keeney and Raiffa 1976, Keeney 1992). If the attributes are additive independent or can be approximated as additive independent, a weighted sum is used. For those who are familiar with differential equations, additive independence is equivalent to separation of variables where a function of multiple variables may be reduced to a linear combination of functions of single variables.

Mathematically, an expected utility, $\hat{u}$, of the form

Table 4-1. ***Consequences and Corresponding Utilities***

Alternative	*Safety % [p(S)]*	*Safety Utility (u_s)*	*Cost (million$)*	*Cost Utility (u_c)*
Flange joint, not tested	99.5	0.97	205	0.00
Flange joint, tested	99	0.94	260	0.00
New tang and clevis	98	0.85	120	0.60
Old tang and clevis	96	0.43	80	0.84

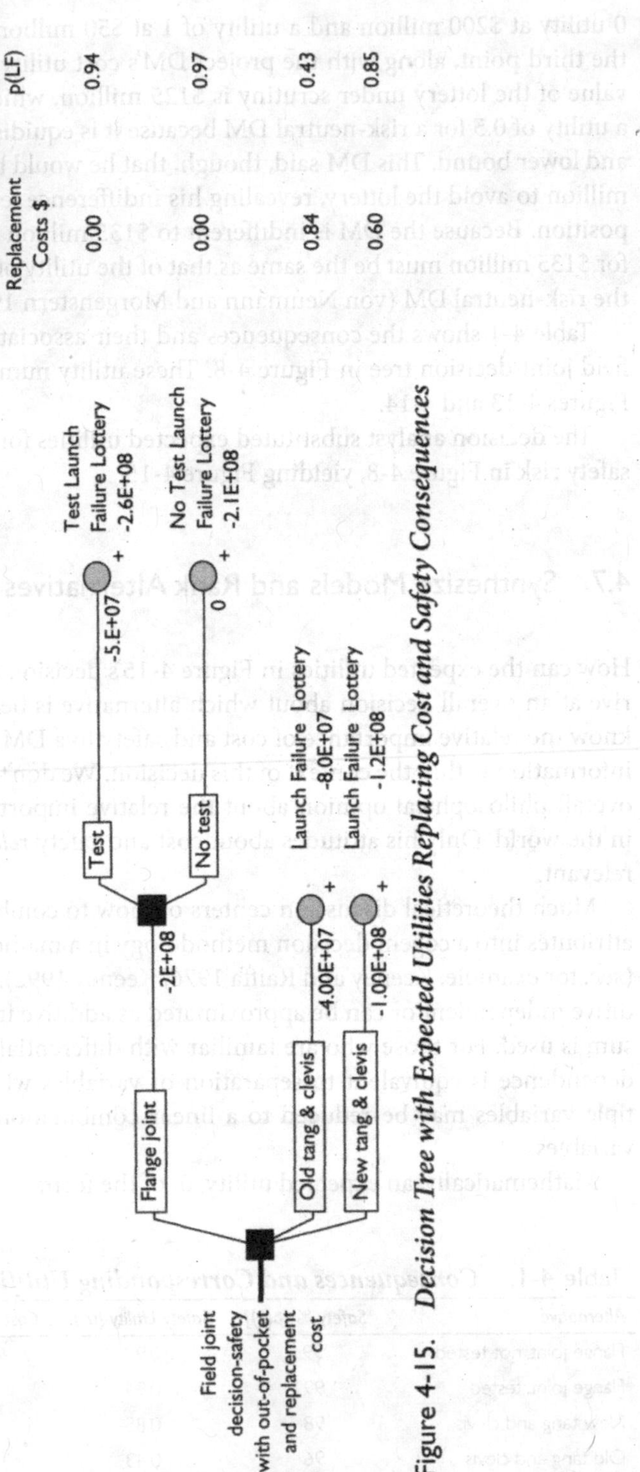

Figure 4-15. ***Decision Tree with Expected Utilities Replacing Cost and Safety Consequences***

$$\hat{u}\,(y_1, \ldots, y_N) = \sum_{i=1}^{N} w_i u_i\,(y_i) \qquad (4\text{-}10)$$

is valid if and only if the attributes y_i are additive independent, where w_i is a constant coefficient for the ith attribute. Analysts typically scale the coefficients such that $\sum_{i=1}^{N} w_i = 1$. In Equation 4-10, N is the number of attributes. Although many situations occur when additive independence does not hold (Keeney and Raiffa 1976), it is often used as an approximation. For a less formal method that does not rely on additive independence, see Chapter 6 of Hammond et al. (1999).

In the case of only two attributes (e.g., in the SRM field joint project), $N = 2$ and the expected utility would be

$$\hat{u}^j = w_s u_s^{\,j} + (1 - w_s) u_c^{\,j} \qquad (4\text{-}11)$$

where

$\hat{u}^j$ = the expected utility of the jth alternative,

w_s = the relative importance of safety,
$u_s^{\,j}$ = the expected utility of safety for the jth alternative, and

$u_c^{\,j}$ = the cost utility for the jth alternative.

Because the weights are normalized to 1, the cost-weighting factor is $1 - w_s$.

Comparing Apples and Oranges?

Everybody makes such comparisons many times every day, doing so intuitively and in seconds. For example, you're late for an important sales meeting and you want to cross a busy avenue in the middle of the block instead of walking (or running) to the corner and waiting for the light. On the one hand is an important career and financial opportunity. On the other hand is your personal safety. Within a second or two, you decide on whether it's worth your personal safety to arrive at the meeting on time. You choose the alternative that you believe maximizes your chance of success and survival at that moment. In the language of utility theory, if you conform to the normative group, you (the DM) select the alternative that maximizes your expected utility.[3]

The recommended alternative would be the one with the largest expected utility such that

$$\text{recommended alternative} = \text{maximum } \{\hat{u}^j\,(y^j_1, \ldots, y^j_N)\} \tag{4-12}$$

where j = 1 to L and L is the number of alternatives.

The individual attribute utilities of each alternative, $\hat{u}^j\,(y^j_i)$, are taken directly from the decision tree. Howard (1988) proved that only the means of consequences are required for a decision. Figures 4-2, 4-6, 4-7, 4-8, and 4-15 use mean quantities of cost and safety. As a result, the utilities shown in Figure 4-15 are the expected utilities for each attribute and each alternative.

The analyst should verify the condition of additive independence for each decision analysis. The condition should hold for that portion of the range of attribute quantities of concern for the decision. It need not hold for all possible values of the attributes (i.e., it need not hold over all possible values of cost and safety) for all future decisions. Let Y_1 and Y_2 be two attributes with values y_{1i} and y_{2j}. Additive independence is proven for a decision problem if it can be shown the DM is indifferent to the following two lotteries:

- Lottery 1: There is a 50% chance that the consequence will be y_{11} and y_{21} and a 50% chance that the consequence will be y_{12} and y_{22}.
- Lottery 2: There is a 50% chance that the consequence will be y_{11} and y_{22} and a 50% chance that the consequence will be y_{12} and y_{21}.

To be strictly mathematically rigorous, this indifference condition should hold for any set of y_{1i} and y_{2j} over the range of magnitudes relevant to the decision.

Returning to the SRM field joint analysis, the DM was asked if he was indifferent between the following lotteries:

- Lottery 1: There is a 50% chance that the cost will be \$75 million and p(S) will be 96% and a 50% chance that the cost will be \$150 million and p(S) will be 99%.
- Lottery 2: There is a 50% chance that the cost will be \$75 million and p(S) will be 99% and a 50% chance that the consequence will be \$150 million and p(S) will be 96%.

The DM reasoned that the first lottery is between low cost, low safety and high cost, high safety, and that the second lottery is between low cost, high safety and high cost, low safety. The worst outcome is in Lottery 2 (high cost, low safety) and the best outcome is also in Lottery 2 (low cost, high safety). The 50/50 lottery means that on the average, the DM would obtain intermediate levels of cost and safety. Similarly, in the first lottery he chose between low cost and safety and high cost and safety. He was reasonably indifferent to these two lotteries.

With the assumption of additive independence for this decision validated, we can think of the coefficients, w_i, as measures of relative importance of one attribute to another. The importance measure should not be regarded as universally applicable. Safety is not always x times more important than cost. The decision analyst uses the importance measures over a range of attribute magnitudes relevant to the decision analysis at hand and, furthermore, only over the range of safety and cost for which the indifference lottery holds.

In principle, the decision analyst would ask the DM a question such as "How much more or less important is safety over cost, if costs are in the approximate range of $70 million to $200 million and for p(S) in the approximate range of 96% to 99%?" Engineering DMs, however, are often reluctant to reveal their attitudes about safety, particularly in comparison with monetary equivalents. This is similar to the reluctance of many people to discuss a cost of human life or cost preferences associated with avoiding health hazards.

To the engineering DM, a useful approach is to parameterize the importance of cost and safety and create a decision trajectory (Frank 1995). A decision trajectory is a set of curves that depict the change in the selection of alternatives as a function of the DM's preferences for safety to cost. Figure 4-16 shows the decision trajectories for each alternative in the SRM field joint analysis. Each line in this figure was developed by substituting the cost and safety utilities from Figure 4-15 into Equation 4-11 and then varying w_s between 0 and 1.

A vertical line drawn through the curves in Figure 4-16 at any point on the x-axis shows the ranking of the alternatives for that w_s based on expected utility. For example, for equal weighting of safety and cost—and based on expected utility—the rankings are

1. New tang and clevis
2. Old tang and clevis
3. Flange joint/not tested
4. Flange joint/tested

Note that this ranking differs from the one derived from the decision model in Figure 4-2. In that ranking, in which the analyst did not explicitly include the DM's preferences and used cost as the consequence of interest, the old tang and clevis design was considered best.

The flange joint alternatives rise in expected utility with safety because the project DM has assigned them 0 utility with respect to cost. Although they are very expensive, they have a safety advantage. Thus, if the DM has a strong bias toward safety (i.e., w_s is larger than approximately 0.85), the untested flange joint alternative becomes the best. If cost is overwhelmingly important from the DM's perspective, the old tang and clevis alternative would be recommended. This holds for w_s below about 0.35. Between these extremes, the new tang and clevis joint is preferred.

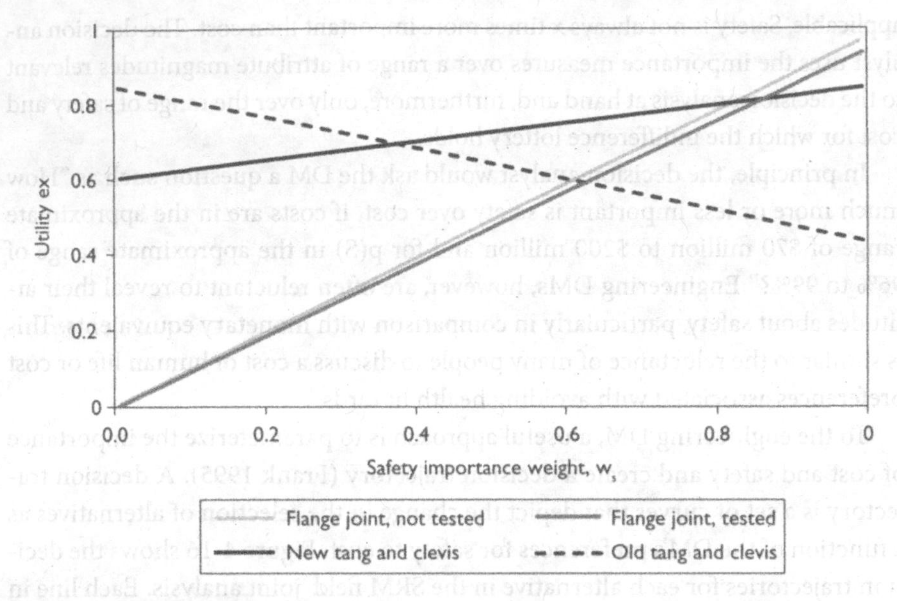

Figure 4-16. ***Resultant Decision Trajectories for Solid Rocket Motor (SRM) Field Joint Analysis***

4.8 Develop a Descriptive Decision Model: The Analytic Hierarchy Process

The second method for developing decision and value models, AHP (Saaty 1990, 2000), is applied using the same general steps as those depicted in Figure 4-1. AHP differs from normative decision analysis in several ways, described next. At the end of this chapter, I summarize some of the scholarly disputes about AHP and MAUT.

Instead of a decision tree, the diagrammatic basis for an AHP is a hierarchy diagram. Figure 4-17 illustrates the general framework of such a diagram. The objective of the decision analysis, which is usually a desire to choose the best alternative, is depicted in the top level of the hierarchy. The bottom level of the diagram depicts the alternatives selected for attaining the objective, and the middle levels delineate the attributes with which alternatives are evaluated. Analysts can subdivide attributes as necessary to obtain a level at which (1) the DM can make judgments and (2) the attribute quantities are measurable or calculable. Figure 4-18[4] is the hierarchy diagram for our example SRM field joint decision.

In essence, the AHP follows these steps:

1. Determine the relative preference or ranking of each alternative at the bottom level with respect to each attribute in the second level. For example, the DM's preferences for each of the joint alternatives are evaluated with respect to cost and then independently evaluated with respect to safety.

2. Determine the relative importance or ranking of each attribute. For example, the DM expresses his values with respect to the relative importance of cost versus safety.

3. Combine steps 2 and 3 to obtain the ranking of each alternative (third level) with respect to the goal (first level).

4.8.1 AHP Step 1: Pairwise Comparisons of Alternatives

To accomplish the first step, a decision analyst constructs matrices whose elements are pairwise comparisons of each alternative for each attribute. As Table 4-2[5] shows, the matrix elements are preferences of the row entry to the column entry. Matrix positions are indicated by (row, column). For example, position (1,4) contains the element "Preference of flange joint, not tested, to old tang and clevis." Taking one cell at a time, a DM answers the question "How much more do I prefer one alternative (the row entry) over the other (the column entry)?" When using AHP, it is customary but not necessary to use an integer greater than or equal to 1 and less than or equal to 9 in answer to such questions. The number is entered into a row, column position of a matrix. It is also customary, but not necessary, to convert qualitative judgments to cardinal integers between 1 and 9.

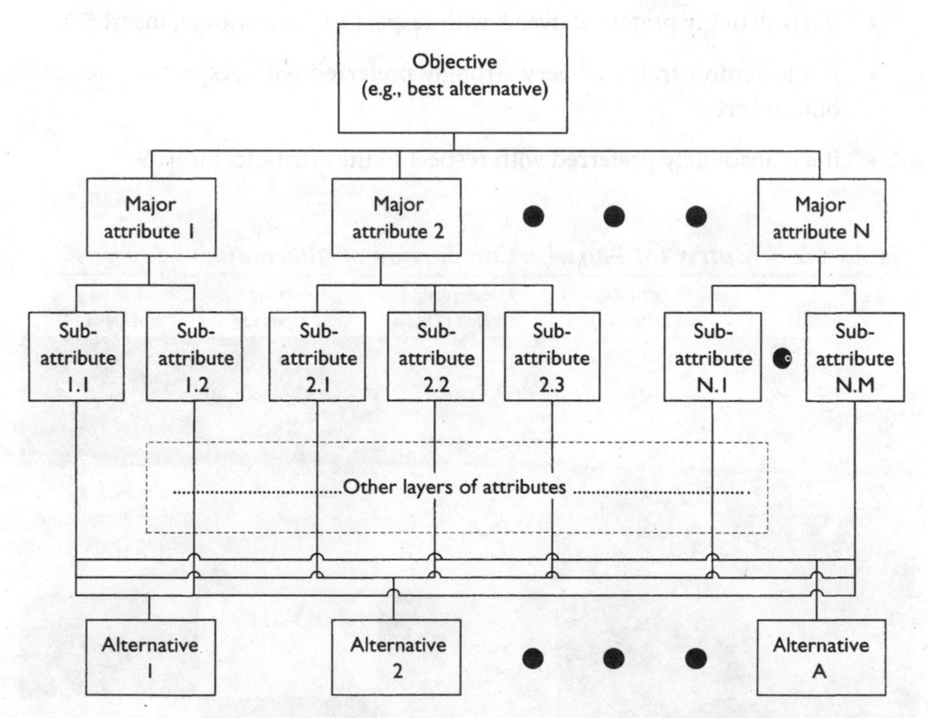

Figure 4-17. ***Framework for Analytic Hierarchy Process (AHP)***

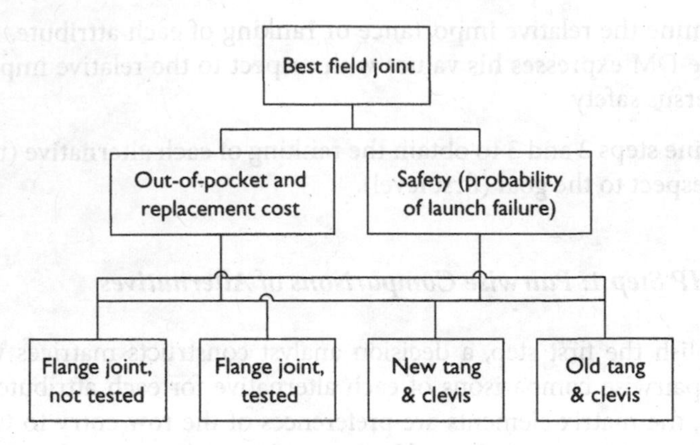

Figure 4-18. ***Hierarchy for Solid Rocket Motor (SRM) Field Joint***

For example, given an element (*a*, *b*) in the matrix, where *a* is the row and *b* is the column:

- if *a* and *b* are equally preferred with respect to the attribute, insert 1
- if *a* is weakly preferred over *b* with respect to the attribute, insert 3
- if *a* is strongly preferred over *b* with respect to the attribute, insert 5
- if *a* is demonstrably or very strongly preferred with respect to the attribute, insert 7
- if *a* is absolutely preferred with respect to the attribute, insert 9.

Table 4-2. ***Matrix for Pairwise Comparison of Alternatives***

Alternatives	*Flange joint, not tested (1)*	*Flange joint, tested (2)*	*New tang and clevis (3)*	*Old tang and clevis (4)*
Flange joint, not tested (1)	1	Preference of flange joint, not tested, to flange joint, tested	Preference of flange joint, not tested, to new tang and clevis	Preference of flange joint, not tested, to old tang and clevis
Flange joint, tested (2)	Reciprocal	1	Preference of flange joint, tested, to new tang and clevis	Preference of flange joint, tested, to old tang and clevis
New tang and clevis (3)	Reciprocal	Reciprocal	1	Preference of new tang and clevis to old tang and clevis
Old tang and clevis (4)	Reciprocal	Reciprocal	Reciprocal	1

The even numbers 2, 4, 6, and 8 are used to facilitate compromise when a DM has difficulty deciding on one of the odd numbers. Because an alternative cannot be preferred more or less than itself, the diagonal elements are always 1. Reflecting the matrix across the diagonal with reciprocal quantities is also logical. Table 4-2 indicates the reciprocal pairs by similar shading. For example, if the untested flange joint is preferred to the old tang and clevis joint—with respect to safety by a ratio of 8 in matrix position (1,4)—matrix position (4,1) should contain the ratio 1/8.

During an AHP, a decision analyst guides the DM through the AHP, helping the DM to understand the significance of each element in the matrices. Furthering our field joint example, the decision analyst developed a matrix such as in Table 4-2 for each attribute. This resulted in two matrices, one for safety and one for cost. The DM made the preference and importance judgments for each element in the matrices shown in Tables 4-3 and 4-4. To develop the element values for Table 4-3, the DM simply used the ratios of the calculated launch failure probabilities developed during the PRA (as previously shown in Figure 4-8).

For example, to fill in matrix position (1,4), the DM divided 4% by 0.5%, indicating that the lower launch failure probability of the flange joint is very strongly preferred. To develop the element ratios for Table 4-4, the DM used the ratios of the calculated costs shown in Figure 4-8. For example, to fill in matrix position (4,1) in Table 4-4, the DM divided \$2.05E+08 by \$8E+07 to obtain 2.56, indicating that the old tang and clevis costs were weakly preferred. This method of filling out the matrices does not include the relative values that the DM placed on expenditures and launch failure probability. Later in this section, I discuss how the DM's values may be included in the AHP using utility functions in a manner analogous to the normative decision model described in Sections 4.6 and 4.7.

Note that the AHP can also be applied without explicit safety or cost calculations. The matrices can be filled out using only the DM's judgments. For important decisions that involve high-consequence outcomes, however, quantitative safety and cost estimates are much preferred as a basis for decision analysis, particularly if demonstrating due diligence in assessing risks is important.

Table 4-3. ***Example Analytic Hierarchy Process (AHP) Safety Matrix***

Alternative	*Flange joint, not tested*	*Flange joint, not tested*	*New tang and clevis*	*Old tang and clevis*
Flange joint, not tested	1	2	4	8
Flange joint, tested	1/2	1	2	4
New tang and clevis	1/4	1/2	1	2
Old tang and clevis	1/8	1/4	1/2	1

Note: Preferences with respect to safety.

Table 4-4. *Example Analytic Hierarchy Process (AHP) Cost Matrix**

Alternative	Flange joint, not tested	Flange joint, tested	New tang and clevis	Old tang and clevis
Flange joint, not tested	1	1.27	0.59	0.39
Flange joint, tested	0.79	1	0.46	0.31
New tang and clevis	1.71	2.17	1	0.67
Old tang and clevis	2.56	3.25	1.5	1

*Preferences with respect to cost.

4.8.2 AHP Step 2: Cost versus Safety Preference

In general, if there are N attributes, the analyst would develop an $N \times N$ matrix within which the DM again makes pairwise comparisons about relative preferences between attributes. In our example, safety and cost are the two attributes, yielding a 2×2 matrix in which the elements were pairwise compared as shown in Table 4-5.

Table 4-5. *Generalized Matrix for Safety and Cost Attribute Importance Weighting*

Attribute	Safety	Cost
Safety	1	Importance of safety relative to cost, η
Cost	Importance of cost relative to safety, 1/η	1

Table 4-6. *Alternative Ranking with Respect to Safety Attribute*

Alternative	Relative preference (eigenvector)
Flange joint, not tested	0.53
Flange joint, tested	0.27
New tang and clevis	0.13
Old tang and clevis	0.07

Table 4-7. *Alternative Ranking with Respect to Cost Attribute*

Alternative	Relative preference (eigenvector)
Flange joint, not tested	0.17
Flange joint, tested	0.13
New tang and clevis	0.28
Old tang and clevis	0.42

4.8.3 AHP Step 3: Develop the Relative Ranking of the Alternatives

Relative Rankings with Respect to Attributes. Thomas L. Saaty's insight was that for any square matrix (i.e., the same number of rows and columns), the elements of the eigenvector with the principal (i.e., largest) eigenvalue of the matrix represent the relative rankings of the alternatives (Saaty 1990). In other words, the overall rankings implied by the pairwise comparisons are found as elements of the eigenvector with the largest eigenvalue. Furthermore, the eigenvalue provides a measure of the consistency of the DM's preferences with respect to transitivity and reciprocity. If the judgments have been entirely logically consistent, the principal eigenvalue will equal the rank of the matrix. In Tables 4-3 and 4-4, the rank of the matrices is 4.

Solving for the principal eigenvectors of the matrix in Table 4-3 and normalizing them to sum to 1 yields a vector whose elements exhibit the relative preference (or ranking) of the alternatives with respect to safety. This is shown in Table 4-6.

Similarly, solving the matrix in Table 4-4 for our SRM field joint example, the cost rankings of the alternatives are given in Table 4-7.

Table 4-7 implies that the old tang and clevis joint was preferred over the flange joint, tested, by 0.423/0.130 = 3.25, which is the same as the input quantity in Table 4-4. The decision analysis concatenated the two attribute eigenvectors in Tables 4-6 and 4-7 to form a 4 × 2 matrix as in Table 4-8.

In Table 4-8, Column 1 represents the safety eigenvector and Column 2 represents the cost eigenvector. Each row is a safety and cost element for an alternative.

Synthesis of Relative Rankings with Attribute Importance. Additive independence of the attributes is assumed in the AHP. Recall that Equations 4-10 and 4-11 combined attribute utilities using a factor, w, that provided the importance of one attribute to another such that the sum of the weighting factors is normalized to 1. The decision analyst in the field joint project used the same approach in the AHP. For example, if the DM believes safety is strongly more important than cost, $\eta = 5$. The matrix of Table 4-5 then looks like that shown in Table 4-9.

This matrix has the principal normalized eigenvector [0.83, 0.17], which means that the safety weighting factor, $w_s = 0.83$ and—just as in Equation 4-11—$w_c = 1 - w_s$.

Matrices and vectors offer a mathematical shorthand for a series of equations of the form of Equation 4-10. The dot product of the 4 × 2 matrix in Table 4-8

Table 4-8. ***Combined Attribute Eigenvector Matrix***

Alternative	*Safety*	*Cost*
Flange joint, not tested	0.53	0.17
Flange joint, tested	0.27	0.13
New tang and clevis	0.13	0.28
Old tang and clevis	0.07	0.42

Table 4-9. *Attribute Weighting Matrix with Safety to Cost Importance = 5*

1	5
1/5	1

with the above eigenvector is shown in Figure 4-19. Performing the dot product of the matrix and eigenvector in Figure 4-19 is equivalent to weighting the preference of each alternative by the relative importance of each attribute. For example, to obtain the normalized relative ranking of the flange joint, not tested, the analyst performed the following:

$$(0.533)(0.83) + (0.165)(0.17) = 0.47$$

which is analogous to Equation 4-11. The normalized vector in Table 4-10 is the result of the dot product procedure. It is the relative ranking of the field joint alternatives that combines both safety and cost attributes.

By varying η between 1/9 (i.e., cost is nine times more important than safety) and 9 (i.e., safety is nine times more important than cost) with repetitive solutions of the eigenvectors and dot products, the decision analyst constructed the decision trajectory as shown in Figure 4-20. For reference, Table 4-11 shows w_s as a function of the safety to cost importance parameter, η.

As safety becomes increasingly important, the options with the highest calculated safety levels emerge with the highest ranking. The field joint decision analysis found that even when safety was considered unimportant relative to cost (i.e., preference of safety to cost < 0.5), the flange joint, not tested, emerged as the best alternative. In this analysis, the calculated costs and calculated safety were used as the basis for preferences without additional value judgments. Safety and cost utility functions were not used in the above AHP example. Next, I expand the same example to include utility functions.

4.8.4 AHP with Utility Functions

Similar but not identical to the normative decision analysis covered in Section 4.7, utility functions in the AHP also substitute for actual calculated consequences when determining the ranking of alternatives (see Section 4.6). Using

Matrix			Eigenvector
Safety	Cost		
0.53	0.17	•	0.83
0.27	0.13		0.17
0.13	0.28		
0.07	0.42		

Figure 4-19. *Matrix and Eigenvector Dot Product*

Table 4-10. *Normalized Rankings of Field Joint Alternatives*

Alternative	*Normalized Eigenvector*
Flange joint, not tested	0.47
Flange joint, tested	0.24
New tang and clevis	0.16
Old tang and clevis	0.13

Note: Combining cost and safety attributes using the analytic hierarchy process (AHP).

the information from the SRM field joint project, the analyst started the process by using the calculated consequences in Figure 4-8 and the DM's utility functions for safety and cost in Figures 4-13 and 4-14. The decision analyst converted the consequences to their equivalent utilities using the relationship between consequences and utilities in these figures. The utility values are all that are needed to create paired comparisons and, thereby, construct a matrix composed of the eigenvectors for each attribute. This process is just like that described in Section 4.8, but instead of using the ratio of consequences from Figure 4-8, the analyst used the ratio of the equivalent utilities.

The remainder of the AHP was applied just like that discussed earlier in Section 4.8. A matrix of eigenvectors was developed from the relative rankings with

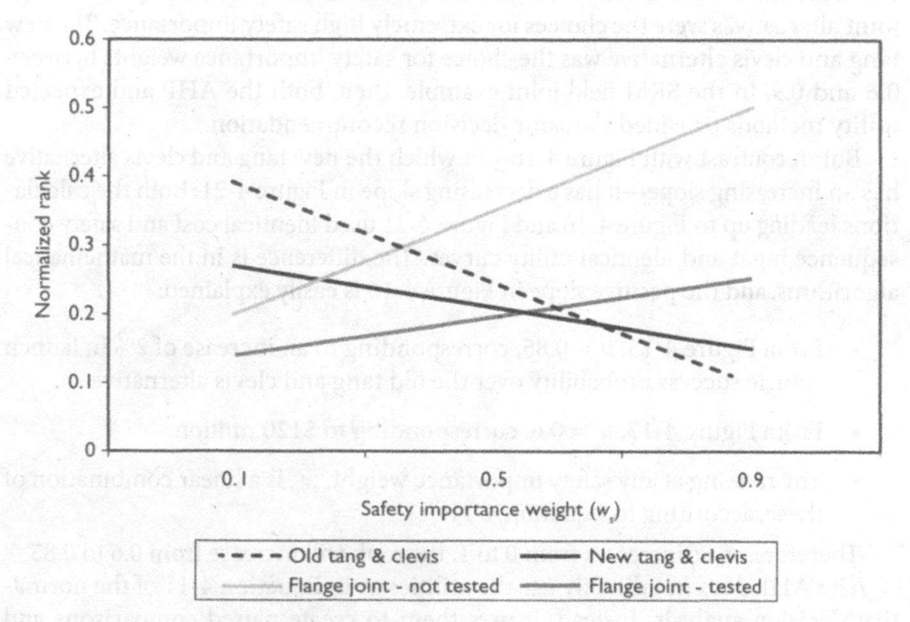

Figure 4-20. ***Safety and Cost Decision Trajectory***

Note: For analytic hierarchy process (AHP) solution to solid rocket motor (SRM) field joint decision.

Table 4-11. ***Safety Weighting Factor, w_s versus Safety to Cost Ratio, η***

	Correspondence between η and w_s						
η	10	5	2	1	0.5	0.2	0.1
w_s	0.909	0.833	0.667	0.5	0.333	0.167	0.091

Table 4-12. ***Combined Attribute Eigenvector Matrix***

Alternative	*Safety*	*Cost*
Flange joint, not tested	0.30	0.0
Flange joint, tested	0.29	0.0
New tang and clevis	0.27	0.41
Old tang and clevis	0.14	0.59

Note: For field joint decision using analytic hierarchy process (AHP) with utility functions.

respect to each attribute (shown in Table 4-12); an eigenvector of the cost versus safety relative importance matrix was developed; and the dot product of the two was taken to obtain the desired alternative rankings. By again varying η, a decision trajectory such as that shown in Figure 4-21 was developed.

In comparing Figure 4-21 with Figure 4-16, we can see that the old tang and clevis alternative has the highest rank for lower safety importance and the flange joint alternatives were the choices for extremely high safety importance. The new tang and clevis alternative was the choice for safety importance weights between 0.6 and 0.9. In the SRM field joint example, then, both the AHP and expected utility methods provided the same decision recommendation.

But in contrast with Figure 4-16—in which the new tang and clevis alternative has an increasing slope—it has a decreasing slope in Figure 4-21. Both the calculations leading up to Figure 4-16 and Figure 4-21 used identical cost and safety consequence input and identical utility curves. The difference is in the mathematical algorithms, and the positive slope in Figure 4-16 is easily explained:

- From Figure 4-15, $u_s = 0.85$, corresponding to an increase of 2% in launch vehicle success probability over the old tang and clevis alternative.
- From Figure 4-15, $u_c = 0.6$, corresponding to \$120 million.
- The ranking at any safety importance weight, w_s, is a linear combination of these, according to Equation 4-11.

Therefore, as w_s increases from 0 to 1, the rank will increase from 0.6 to 0.85.

The AHP does not directly use the utilities as in Equation 4-11 of the normative decision analysis. Instead, it uses them to create paired comparisons and then a normalized eigenvector, as I described in Section 4.8.3. For the new tang and clevis alternative, the normalized cost rank from this process is 0.41 and the normalized safety rank is 0.27 (as shown in Table 4-12). These are linearly

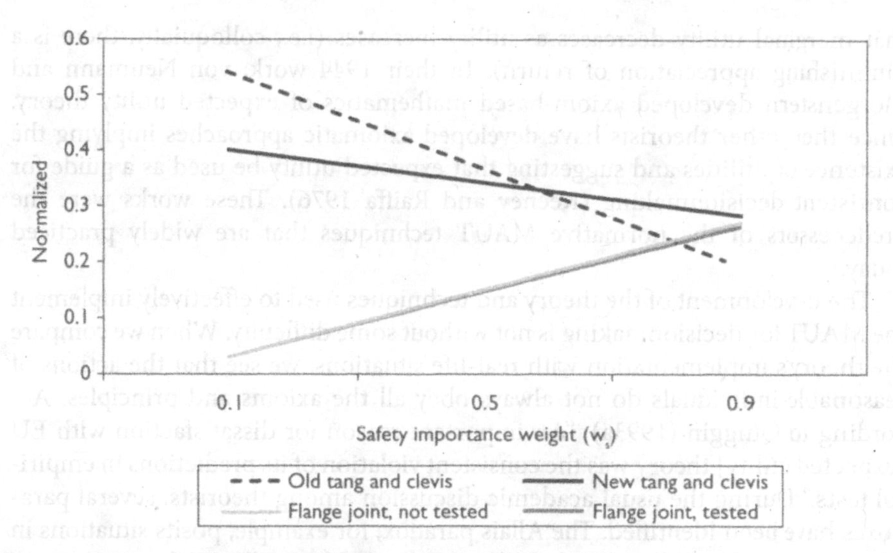

Figure 4-21. ***Safety and Cost Decision Trajectory***

Note: For analytic hierarchy process (AHP) utility function solution to solid rocket motor (SRM) field joint decision.

combined with w_s as part of the dot product of the attribute matrix and the weight eigenvector (see Figure 4-19). As w_s increases from 0 to 1, the rank will decrease from 0.41 to 0.27.

The AHP and normative utility theory, therefore, will not reproduce each other's quantitative results.

Curious about what really happened during the SRM field joint project? The DM terminated the flange joint development program in favor of a new tang and clevis design.

4.9 Controversies Surrounding AHP and MAUT

The notion that people make decisions not on the expected *value* of an endeavor but by their expected *utility* had its origin with Daniel Bernoulli in 1738, when he proposed a solution to the St. Petersburg paradox. The paradox posed the following situation—a fair coin is tossed until a head appears; if the first head appears on the nth toss, the payoff is 2^n ducats. The paradox is that the expected return of this game is infinite. According to the expected value theory during Bernoulli's time, a person should be willing to pay just slightly less than infinite to play. Because very few people would pay such a price, the entire notion that people make decisions on the expected return of an endeavor was called into question. Bernoulli proposed an alternative hypothesis that a person's valuation of a venture (or game) is the expected utility, not the expected return, and that marginal utility decreases as utility increases (i.e., colloquially, there is a diminishing appreciation of return). In their 1944 work, von Neumann and Morgenstern developed axiom-based mathematics of expected utility theory. Since then other theorists have developed axiomatic approaches implying the existence of utilities and suggesting that expected utility be used as a guide for consistent decisionmaking (Keeney and Raiffa 1976). These works were the predecessors of the normative MAUT techniques that are widely practiced today.

The development of the theory and techniques used to effectively implement the MAUT for decisionmaking is not without some difficulty. When we compare the theory's implementation with real-life situations, we see that the actions of reasonable individuals do not always obey all the axioms and principles. According to Quiggin (1993a), "An important reason for dissatisfaction with EU [expected utility] theory was the consistent violation of its predictions in empirical tests." During the usual academic discussion among theorists, several paradoxes have been identified. The Allais paradox, for example, posits situations in which DMs, when offered a choice between gambles, will not choose them in accordance with the traditional principle that a DM should be indifferent to two lotteries with the same outcome (Allais 1953). Eventually, theorists made considerable progress in addressing this problem using *prospect theory, cumulative prospect theory,* and *rank dependent expected utility theory* (Kahneman and Tversky 1979; Tversky and Kahneman 1992; Quiggin 1993b). The Ellsberg paradox posits situations in which DMs will choose a lottery that has a lower expected utility than another lottery of equal or larger expected utility when the outcome of the larger is more ambiguous (i.e., has higher epistemic uncertainty; Ellsberg 1961). This notion of ambiguity aversion is not accounted for in expected utility theory, but some progress has been made in demonstrating limitations in the applicability of this paradox (see, for example, Fox and Tversky 1995). Other paradoxes have been offered and comparisons of empirical situations with utility theory continue, as does development of the theory.

Against the backdrop of this history, Thomas L. Saaty introduced the AHP (Saaty 1980). It too rests on the notion of expected utility but arrives at such preferences in an entirely different way. Both MAUT and AHP are multiattribute decision analysis methods, and an academic controversy continues to this day about the relative merits of each method. I don't intend to adjudicate or resolve this controversy here, nor do I discuss the arguments in detail. Other sources have detailed the controversies (see, for example, Triantaphyllou 2000). Practitioners of the methods I describe in this book, however, should find understanding the key elements of the controversy useful. Part of the controversy is framed around strict transitivity and cardinal preferences, which characterize the normative model of expected utility theory, and ordinal preferences and rank reversal, which characterize the descriptive model of AHP. Strict transitivity means that if A is preferred to B and B is preferred to C, then A must be preferred to C. This is a fundamental axiom of utility theory and appears self-evident to many people. AHP does not make use of this axiom and has been criticized as non-normative. But in many counterexamples and in empirical evidence, researchers have identified instances of perfectly reasonable people making reasonable decisions that violate transitivity (see, for example, Li 1996). The AHP provides a measure of deviation from strict transitivity as a guide to the decision analysis (Saaty 1990).

The AHP uses the eigenvector or the principal eigenvalue of a matrix to extract alternative rankings. Considerable debate continues about the validity of the eigenvector method with Saaty's preference scale. Barzilai (2001), for example, argues that a ratio scale implies that the property of the variable expressed as a ratio must have an absolute zero. He goes on to say that an eigenvector method is admissible for ratio scales but not ordinal scales with no underlying absolute zero. Saaty (2005), on the other hand, argues that the absolute zero requirement is not needed. An analyst, therefore, should take care in defining preferences when using the AHP. In the examples in this book, AHP is applied using costs in terms of dollars, safety in terms of probabilities of events or end states, and reliability in terms of probability of success. In this book, I have used AHP to establish preferences using ratios of these, each of which has an absolute zero.

Normative theory also holds that addition of irrelevant alternatives must not change the rankings of the original alternatives (Luce and Raiffa 1957). The original 1980 version of the AHP was criticized as allowing this behavior, called rank reversal, to occur. Some theorists who advocate MAUT argue that a method in which such rank reversal occurs is fundamentally flawed and cannot be trusted to make the right recommended decision (Belton and Gear 1982; Dyer 1990). Recent work, however, has shown that other widely used and long-standing methods (e.g., ELECTRE and TOPSIS; see Table 4-13) also undergo rank reversal (Wang and Triantaphyllou 2006). Many counterexamples, both in real-life situations and devised by theorists, demonstrate that under some circumstances, rank reversal is proper and does occur if additional alternatives are introduced (Saaty 2000). To respond to this criticism, modifications to the original AHP were developed to ensure that rank reversal would not occur (Barzilai and Golany 1994) or to limit the situations in which it would (Saaty and Vargas 1984). Currently, theorists believe that the weighted product model (WPM; see Bridgeman 1922 and Miller and Starr 1969) is immune from rank reversal.

Rank-reversal difficulties with the AHP have also been traced to the use of the dot product of the preference matrix elements and attribute weights. This is an additive expression because the preferred alternative is the one that maximizes the sum of the products of the preference matrix elements and attribute weights. Barzilai and Lootsma (1997) and Lootsma (1999) proposed a multiplicative version of the AHP method. According to this method, the relative performance ratios and criteria weights are not processed according to an additive expression. Instead they are processed according to a multiplicative expression as first proposed in the WPM (Bridgeman 1922; Miller and Starr 1969). Triantaphyllou (2000) proved that most of the ranking irregularities that occurred when the additive variants of the AHP method were used will not occur with the multiplicative AHP method.

Table 4-13. ***A Selection of Other Multiattribute Decision Analysis Methods***

Method	Original Reference
SMART	von Winterfeldt and Edwards (1986)
Value-Focused Thinking	Keeney (1992)
PROACT	Hammond et al. (1999)
ELECTRE III	Roy (1978)
Technique for Order Preference by Similarity to Ideal Solution (TOPSIS)	Hwang and Yoon (1981)
Weighted Product Model (WPM)	Miller and Starr (1969); Bridgeman (1922)

I neither advocate nor disparage the AHP and MAUT. None of the examples and case studies in this book, all of which are practical decision situations related to safety or reliability, added new alternatives after the decision analysis had been completed. Rank reversal, therefore, did not occur. This chapter solves the same decision problem using both methods as does the case study in Chapter 7. And, not surprisingly, different rankings can be obtained using different methods. Decision theorists continue to study the issue of different accepted and widely practiced methods giving different rankings. The key question is, of course, which one yields the more reliable decision recommendation.

From a practitioner's viewpoint, both methods have been studied, criticized, and compared to empirical results and to each other. The methods continue to evolve to increase agreement between real behavior and theory, remove theoretical objections, and increase their applicability to an ever-widening scope of decision problems. Both methods work to aid DMs in most practical situations. Many of the methodology anomalies have resulted from devised situations or studies using randomized inputs. Selecting a method is often a matter of judgment and convenience to suit the problem, as long as the decision analyst understands the method's limitations. For example, Chapter 10 obtains expected utilities of alternative safety strategies using a decision tree (i.e., a normative process) because the quantitative evaluation of chance nodes following a decision node about alternative strategies seemed "natural" to the researchers who were familiar with event trees in PRA. On the other hand, the case study in Chapter 6 describes a successful use of AHP to help make a consistent and satisfying decision about which wind tunnel compressor blade option to choose. Knowing the controversies about the methods helps the practitioner avoid the pitfalls of either method. Experienced decision analysts test their methods by performing sensitivity studies on the rankings. As I do more than once in this book, using more than one method to perform the analysis may prove useful. If the recommended alternative is the same for both methods, as in the field joint example in this chapter, the analyst and the DM can have more confidence in the results. Differences in the results should be explained, as was also done in the field joint example.

I focus this book on two widely used multiattribute decision analysis methods: MAUT and AHP, but other accepted methods are presented in Table 4-13 with references for further reading.

Notes

1. A high-consequence system or product is one whose failure can cause great harm, injury, or death.
2. In practice, these assumptions may not hold over the entire range of attribute magnitudes. We can, however, develop different value models for different attribute magnitude ranges in such a way that the assumptions approximately hold over each range.
3. See the discussion in Section 4.9 about situations in which DMs do not conform to this maximum utility behavior.
4. Figures 4-18 through 4-21 refer to our SRM field joint example.
5. Tables 4-2 through 4-12 refer to our SRM field joint example.

References

Allais, M. 1953. Le comportement de l'homme rationnel devant le risque, critique des postulats et axiomes de l'école Américaine. *Econometrica* 21: 503–546.

Barzilai, J. 2001. Notes on the Analytic Hierarchy Process. *Proceedings of the NSF Design and Manufacturing Research Conference*, Tampa, Florida, 1–6. Available at http://myweb.dal.ca/barzilai/papers/AHP%20Notes%202001.pdf (accessed June 9, 2007).

Barzilai, J., and B. Golany. 1994. AHP Rank Reversal, Normalization and Aggregation Rules. *INFOR* 3: 57–64.

Barzilai, J., and F.A. Lootsma. 1997. Power Relations and Group Aggregation in the Multiplicative AHP and SMART. *Journal of Multi-Criteria Decision Analysis* (6)3: 155–165.

Bedford, T., and R. Cooke. 2001. *Probabilistic Risk Assessment: Foundations and Methods.* Cambridge, UK: Cambridge University Press.

Belton, V., and T. Gear. 1982. On a Shortcoming of Saaty's Method of Analytic Hierarchies. *Omega* 11(3): 228–230.

Bridgeman, P.W. 1922. *Dimensionless Analysis.* New Haven, CT: Yale University Press.

DoD (Department of Defense). 1980. Military Standard, Procedures for Performing a Failure Mode, Effects, and Criticality Analysis. Mil-Std-1629A. Washington, DC: DoD.

Dyer, J.S. 1990. Remarks on the Analytic Hierarchy Process. *Management Science* 36(3): 249–258.

Ellsberg, D. 1961. Risk, Ambiguity, and the Savage Axioms. *Quarterly Journal of Economics* 75: 643–669.

Fox, C.R., and A. Tversky. 1995. Ambiguity Aversion and Comparative Ignorance. *Quarterly Journal of Economics* 110(2): 585–603.

Frank, M. 1995. Choosing Among Safety Improvement Strategies: A Discussion with Example of Risk Assessment and Multi-Criteria Decision Approaches for NASA. *Reliability Engineering and System Safety* 49(3): 311–324.

Hammond, J., R. Keeney, and H. Raiffa. 1999. *Smart Choices: A Practical Guide to Making Better Decisions.* Boston: Harvard Business School Press.

Howard, R. 1988. Uncertainty about Probability: A Decision Analysis Perspective. *Risk Assessment* 8(1): 91–98.

Hwang, C.L., and K. Yoon. 1981. *Multiple Attribute Decision Making: Methods and Applications.* Berlin: Springer-Verlag.

Kahneman, D., and A. Tversky. 1979. Prospect Theory: An Analysis of Decision under Risk. *Econometrica* XLVII: 263–291.

Keeney, R. 1992. *Value-Focused Thinking: A Path to Creative Decision-making.* Cambridge, MA: Harvard University Press.

Keeney, R., and H. Raiffa. 1976. *Decisions with Multiple Objectives: Preferences and Value Trade-offs.* New York: Cambridge University Press.

Kirkwood, C. 1991. Notes on Attitude Toward Risk Taking and the Exponential Utility Function. http://www.public.asu.edu/~kirkwood/DAStuff/refs/risk.pdf (accessed June 9, 2007).

Kirkwood, E. 2002. Decision Tree Primer. www.public.asu.edu/~kirkwood/DAStuff/decisiontrees/DecisionTreePrimer-2.pdf (accessed June 9, 2007).

Li, S. 1996. An Additional Violation of Transitivity and Independence Between Alternatives. *Journal of Economic Psychology* 17: 645–650.

Lootsma, F.A. 1999. *Multi-Criteria Decision Analysis via Ratio and Difference Judgment (Applied Optimization).* Dordrecht, the Netherlands: Kluwer Academic Publishers.

Luce, R.D., and H. Raiffa. 1957. *Games and Decisions.* New York: John Wiley & Sons.

McCool, A.A. 1991. Space Shuttle Solid Rocket Motor Program, Lessons Learned. AIAA Paper #91-2291-CP. Atlanta, GA: George C. Marshall Space Flight Center.

Miller, D.W., and M.K. Starr. 1969. *Executive Decisions and Operations Research.* Englewood Cliffs, NJ: Prentice-Hall, Inc.

Quiggin, J. 1993a. Testing Between Alternative Models of Choice Under Uncertainty—Comment. *Journal of Risk and Uncertainty* 6(2): 161–164.

———. 1993b. *Generalized Expected Utility Theory: The Rank-Dependent Model.* Boston: Kluwer Academic Publishers.

Roy, B. 1978. Electre III: Un algorithme de classements fondé sur une représentation floue des préférences en présence de critères multiples. *Cahiers du Centre d'Etudes de Recherche Opérationnelle* (Belgique) 20(1): 3–24.

Saaty, T. 1990. *Multicriteria Decision Making: The Analytic Hierarchy Process.* AHP Series, Vol. 1. Pittsburgh, PA: RWS Publications.

———. 2000. *Fundamentals of Decision Making and Priority Theory with the Analytic Hierarchy Process.* AHP Series, Vol. 6. Pittsburgh, PA: RWS Publications.

———. 2005. Making and Validating Complex Decisions with the AHP/ANP. *Journal of Systems Science and Systems Engineering* 14(1) March: 1–36.

Saaty, T.L. 1980. *The Analytic Hierarchy Process.* New York: McGraw-Hill.

Saaty, T.L., and L.G. Vargas. 1984. Inconsistency and Rank Preservation. *Journal of Mathematical Psychology* 28(2): 205–214.

Schmid, S. 1994. The ESA Risk Management Decision Analysis Method: Application Procedure. Report prepared under European Space Research and Technology Center contract, Technology International Corporation. December.

Schuyler, J. 2001. *Risk and Decision Analysis in Projects*, 2nd ed. Newton Square, PA: Project Management Institute.

Shachter, R. 1988. Probabilistic Inference and Influence Diagrams. *Operations Research* 36: 589–604.

Triantaphyllou, E. 2000. *Multi-Criteria Decision Making Methods: A Comparative Study.* Boston: Kluwer Academic Publishers.

Tversky, A., and D. Kahneman. 1992. Advances in Prospect Theory: Cumulative Representation of Uncertainty. *Journal of Risk and Uncertainty* 5: 297–323.

von Neumann, J., and O. Morgenstern. 1944. *The Theory of Games and Economic Behavior*, 2nd ed. Princeton, NJ: Princeton University Press.

von Winterfeldt, D., and W. Edwards. 1986. *Decision Analysis and Behavioral Research.* Cambridge, UK: Cambridge University Press.

Wang, X., and E. Triantaphyllou. 2006. Ranking Irregularities When Evaluating Alternatives by Using Some Multi-Criteria Decision Analysis Methods. *Handbook of Industrial and Systems Engineering.* Edited by A. Badiru. Boca Raton, FL: CRC Press.

CHAPTER 5

Principles of Risk Communication within a Project

THE FUNDAMENTAL REASON FOR performing risk assessment is that it is useful for making decisions in the face of uncertainty about the future. Risk assessment benefits someone who needs to make decisions to meet an organization's or project's objectives. Few schools teach probabilistic risk assessment (PRA), and it is not a discipline normally associated with engineering and management. The ability to communicate methods, results, and benefits of risk assessment is even further removed from the normal education and experience of an engineer. As a group, decisionmakers (DMs) are often uninformed about risk assessment methods and benefits. It is up to the risk analysts to make the case for the usefulness of risk analyses on a project. The ability to clearly communicate risk issues, methods, and results with a high degree of credibility and in a way that is obviously targeted toward the overall success of the project becomes at least as important as the analysis itself.

Sometimes, the best of analyses go unheeded and unused because the results and their significance to the project or organization have not been properly explained. With the realization that DMs in an organization may be as uninformed about quantitative risk methods as lay members of the public, communication with a DM can be structured in a manner similar to communication about risk with the public. There is a long history of research into this area.[1] The principles associated with successful communication about risk to lay people in the public domain can be adapted for use within an engineering or manufacturing organization.

Gaining acceptance of quantitative risk methods within an organization may be aided more by excellent communication abilities than excellent analytical abilities. I don't mean that analysis is unimportant; clearly, competent engineering is critical. But the lack of clear communication with the objective of the organization and project clearly in mind allows those who may not fully understand the methods to perceive the analysis as useless. Successful communication of risk information within an organization involves a change in the attitude of engineers, managers, and DMs. Usually, the burden of educating management about the reasons for doing risk assessment and clearly communicating the results falls on the engineers.

5.1 Principles of Good Communication

Communicating with project team members within a technical organization is much like communicating with the public. Managers, engineers, scientists, and DMs, who are not trained in risk management and thus are laypersons in this area, understand and react to statements about risk in an uninformed manner just like members of the public. In this section, I present 11 principles, based on my own experience, that engineers can use to communicate risk information to DMs.

Principle 1: Adopt the DM's attitude and way of thinking. Like members of the public, a DM sees risk as something different than the accumulation of the results of statistics and computer models. In many ways, a DM's reaction to risk information is predictable. For example, the suggestion (e.g., from a risk management study) that the budget might be overspent is likely to produce an overreaction, while the conclusion that one or more system components might be unreliable will be greeted with apathy. One reason for this is that risk perception is not linear. Some risk-averse DMs will pay more to protect against a loss of performance (even a low probability of such loss) than to pursue a substantial gain in performance. From the DM's perspective, this is actually reasonable because contracts are usually written with substantial penalties for underperformance and few incentives for overachievement.

To effectively communicate, then, engineers and risk analysts need to understand and adopt the DM's attitudes and way of thinking. It's too easy for a risk analyst to say that the boss needs educating. Instead, the risk analyst should strive to learn about the DM's attitudes, his needs, and his ways of thinking about risk. This education process is often influenced by the company's culture as well as the DM's relationship with peers and employees. In moving toward understanding the DM, the analyst needs to get answers to questions such as

- Does the DM trust or distrust his fellow workers?
- What was his previous experience with risk-management activities, if any?
- Is the company willing to invest for a potential future gain and at what probability of success?
- Does the company want to be perceived as progressive or "green"?
- Is there a strong corporate culture of morality to do the right things?
- How far will the company go to avoid appearing to put a cash value on human life?

Principle 2: Minimize technical terminology and discussion of methodology. Using technical jargon when communicating risk-management results is counterproductive. Each scientific and engineering discipline develops its own set of technical terminology. Risk assessment is no exception. Speaking a common language (1) identifies the speaker as a member of the same technical community, (2) allows communication to take place more rapidly among members of the same technical community, and (3) assumes that the listener is well versed in the jargon. But a DM doesn't need to be and most often is not part of the risk assessment technical community. He doesn't understand the technical terminology, and forcing him to ask about each phrase used—or worse yet, forcing him to make an incorrect assumption—is counterproductive. For example, in safety jargon, the term "redundant" indicates a strategy to increase the number of ways to perform a function. In normal English usage, however, redundant means superfluous or excessive.

Engineers tend to communicate their work as if giving a chronological record of what they did and how they did it. But, for a DM, this usually proves too technical, confusing, and long-winded. The DM needs only to hear (briefly) what was done, what was learned (i.e., the conclusions), and what is recommended, all in a way he understands. In making a lengthy presentation of methodology, the focus of the presentation—which should be the results and insights helpful to making the decision—is lost.

Principle 3: Use positive phrases. Like members of the public, DMs are susceptible to the style of verbal presentation of risk information. For example, a 30% chance of failure on the next test of a system sounds worse than a 70% chance of success. A proposal to spend $100,000 to "reduce risk" sounds much worse than a proposal that says the same risk reduction can be accomplished with only 0.1% of the budget. How DMs perceive risk influences their reaction to a risk-management program.

Principle 4: Don't tell a DM what to do. In communicating with DMs, control is a central issue. A project manager is responsible for seeing that project successfully to completion. The risk analyst is not. The DM must retain the ability and authority to decide on tolerable levels of risk. She will resist attempts or suggestions, however oblique, that the analyst should assume a more active decisionmaking role. Even if the DM and the analyst come to the same conclusion, the DM may perceive that the analyst is trying to usurp control. Because this makes the DM uncomfortable, even the *perception* that her control will be diluted by a risk-management or risk-reduction program should be avoided.

Principle 5: Motivate the DM. Like any member of the public, a DM perceives risk through the filter of his own background of experience and needs. Remember, to the DM, risk is not just the accumulation of statistical or predictive models. It may appear to the analyst that the recommendations resulting from the risk-management work represent the only logical and reasonable course of action. The DM, though, doesn't live in the idealized world of risk assessment. His broader concerns must be respected. For example, the DM is not a risk expert, but he is an expert on what he's worried about with respect to his project or company. If these aren't factored in some way into the risk-management process, the DM typically won't accept the results. The DM will be more motivated to accept a risk-management program if it addresses his worries and helps him with his problems.

Principle 6: Bring the DM into the process. From a different perspective, control underlies much of a DM's reluctance to embrace a risk-management program for two specific reasons. First, she usually doesn't understand the process and its underlying technologies. Second, she doesn't trust that anyone else is looking after her interests as diligently as she is.

Bringing the DM into the process is one ingredient of successfully implementing a risk-management program. The analyst shouldn't simply present analysis results as a *fait accompli* after it's all done. Instead, the analyst should involve the DM at the beginning in setting the objectives, scope, and level of effort; in the middle by making mid-course corrections to scope and objectives; and at the end where she can suggest exactly what form of the information she'll find most useful. In this way, communication is taking place throughout the process. This helps the DM to trust in the process and in the analysts themselves and allows her to develop confidence that everyone is working toward the same goal.

Principle 7: Start small. One proven way for the DM to gain confidence in the analyst and the risk-management process is to succeed in a series of small "pilot" projects. Instead of proposing to risk manage the entire project, the risk analysts should consider proposing small test cases for which a clear benefit is easy to see. For example, a project may be investigating alternative valve designs for liquid rocket engines. Here a reliability, safety, cost, performance and schedule risk assessment would yield pointed, useful information for deciding on the best design. Success with each pilot project not only increases the DM's confidence, but also allows the analyst to learn about the DM's attitudes and needs.

Principle 8: Acknowledge the validity of questions and comments. In all communications with the DM, the risk analyst should acknowledge that the DM's questions are valid and sensible, indicate that they are understood, and answer those questions with the specific information requested. If the risk analyst has prepared well by adopting the DM's attitude and point of view, the analyst will be able to anticipate the questions.

Principle 9: Build confidence in the results. The DM needs reassurance that the analysis followed a reasonable process and that the results make sense. To reach this goal, the analyst must display the results as if they were common sense, even if the analysis was quite sophisticated. This is not intended to demean or trivialize the work, but experience is unambiguous here. It's far less effective to try to impress the DM with analytical prowess than it is to impress him with a simple and clear presentation of results and conclusions.

Principle 10: Remain confident but not arrogant. Appearing confident during a presentation is important, but being pedantic, patronizing, or overly authoritative reduces the presentation's appeal. Because the DM may not understand the details, she will rely on her judgment of the presenter as trustworthy and as an expert in the field. She'll judge the presenter's expertise along with the worth of the results to her objectives. The DM will call on her experience to judge if the analyst really knows risk assessment, understands the project's objectives, comprehends the organization's culture, and is acting in the best interests of the organization or project. The risk analyst should point out the assumptions and uncertainties, including the limitations of applicability. This increases the presentation's credibility because DMs know that there are always limits to analysis. This also lays the groundwork for changes in results as new information arises during the project.

Principle 11: Help the DM do his job. Risk analysts shouldn't try to make the DM understand their job. It's not necessary for an experienced DM to understand the details of how risk analysts perform their work, and an attempt at presenting those details will only waste time and possibly confuse him. He needs to know the information that will help him make the decision. The risk analyst, drawing on the understanding of the DM's job gained throughout the project, should present only the risk results and the insights that will help him do that job.

Presenting the quantitative information is just the first step in presenting results. The DM has a right to and usually wants to know what the analyst thinks about the results, reasoning that a risk analyst probably has perspectives that he doesn't possess. Stating a well-timed key insight while sitting at the conference table with the DMs is often more effective for communicating the worth of an analysis than a detailed analysis presentation.

Successful communication of any kind depends on establishing a relationship of mutual respect and trust among parties. Risk analysts, engineers, and DMs are no exception. This is best achieved if the analyst can adopt the DM's perspective following the principles I give in this chapter.

Note

1. See, for example, Sandman (1986); Covello and Allen (1988); NRC (1989); Morgan (1993); Fisher et al. (1995).

References

Covello, V.T., and F. Allen. 1988. *Seven Cardinal Rules of Risk Communication.* Washington, DC: U.S. Environmental Protection Agency.

Fisher, A.M., S. Emani, and M.T. Zint. 1995. *Risk Communication for Industry Practitioners: An Annotated Bibliography.* McLean, VA: Society of Risk Assessment, Risk Communication Specialty Group.

Morgan, M.G. 1993. Risk Assessment and Management. *Scientific American* (July): 40–41.

NRC (National Research Council). 1989. *Improving Risk Communication.* Washington, DC: National Academy Press.

Sandman, P.M. 1986. *Explaining Environmental Risk.* Washington, DC: U.S. Environmental Protection Agency.

CHAPTER 6

The Blade-Trade Case Study

BUILT IN THE EARLY 1950s at NASA Ames Research Center in Mountain View, California, the Unitary Plan Wind Tunnel (UPWT) is now a National Historic Landmark. It was a research facility used extensively to design and test new generations of aircraft, both commercial and military, as well as NASA space vehicles, including the space shuttle. The UPWT is a set of three wind tunnels that comprise three test sections: an 11 × 11 ft transonic tunnel (Mach 0.40 to 1.40); a 9 × 7 ft supersonic tunnel (Mach 1.55 to 2.50); and an 8 × 7 ft supersonic tunnel (Mach 2.45 to 3.45), all capable of operating at variable stagnation pressures. The major common element of the tunnel complex is its drive system, which consists of four interconnected electric motors that can provide 134.23 MW (180,000 hp) continuously. The transonic wind tunnel is a closed-return, variable-density tunnel with a fixed geometry, a ventilated throat, and a single-jack flexible nozzle. Airflow is produced by a three-stage, axial-flow compressor. The original compressor contained aluminum blades.

Accidents that damage wind tunnels have two general causes: human error, in which parts are left in the tunnel, and design problems. A compressor blade, fragmented by either cause, can accelerate to transonic speeds and cause great damage to the wind tunnel or the aircraft or spacecraft model that is being tested.

During the 1980s and early 1990s, the UPWT underwent a major refurbishment and upgrade. As part of that upgrade, designers and engineers considered modifying the compressor and blades of the transonic wind tunnel[1] to enhance performance, minimize operating costs, and improve safety.

In the following sections, I describe how the decisions were reached during the UPWT modification.

6.1 The Decision Opportunity

The original compressor and blade design was such that the ramp up to operational speed went through several resonance frequencies. The project team was considering the modifications not only to eliminate these operationally annoying resonances, but also because of wear and tear on the aluminum blades, which become pitted because of hits from high-velocity fragments and debris in the tunnel. Changing out pitted blades and grinding them to remove the pits was contributing a substantial amount to operating costs. Finally, the potential existed for blade failures to damage the wind tunnel.

Several of the engineering managers wanted to purchase new compressor blades but wanted to make sure that this was justifiable. Collectively, they identified an informal list of factors that represented the areas of concern and the hoped-for improvements resulting from purchasing new compressor blades. The list of areas and concerns included blade life, blade/hardware replacement cost, life-cycle cost (LCC), lost productivity time, energy usage, maximum Mach number capability, necessary management attention, catastrophic failure probability, catastrophic failure damage, and recovery period after catastrophic failure. The project manager (PM) needed a way to pull together all these factors into a cogent, justified recommendation for a course of action. A decision analysis seemed to be the natural way to go.

6.2 The Problem Statement

The PM's perspective was somewhat broader than that of the engineers. In effect, he wanted to ascertain whether the modifications would ameliorate the concerns and, if so, if the benefits would outweigh the expense.

6.3 The Objectives and the Attributes

The decision analyst's task was to identify alternatives that would ameliorate the concerns and then recommend an alternative. The analyst could not use the factors in Section 6.1 as decision attributes because they are not independent. The concerns, then, must be combined in a way that produces independent attributes. LCC includes, for example, blade replacement cost, blade life, energy usage, and lost productivity. The total amount of damage during a time interval depends on the severity of the damage and the frequency of such events. The decision analyst combined these two common components into the safety risk of catastrophic damage and decided to treat the Mach number as a constraint. In other words, an alternative that did not meet the minimum Mach number performance requirement would be disqualified from consideration. Finally, management attention could be lessened if both LCC and safety risk were reduced. As a result, the PM concluded that the decision analysis would be based on two attributes: LCC and safety risk. Separation of safety risk from LCC arose from the consensus that people often consider safety to be something more than equivalent costs. In other words, intangible elements are associated with safety, and these would make safety more important than the expected value of lost revenue from catastrophic failures would indicate. Therefore, for each alternative, the decision analyst, with the help of a PRA analyst, calculated LCC with uncertainties and catastrophic blade damage frequency with uncertainties.

The PM wanted to make a decision based on a risk-neutral perspective. Accordingly, the analyst decided to rank the alternatives without introducing utility functions. This led to a restatement of the objective, which was now to identify the alternative with the best combination of LCC and safety.

6.4 The Alternatives

With the help of engineers on the project, the decision analyst arrived at four alternatives, described in the sections that follow.

Alternative 1: Status quo. This alternative represented the continued use of the current aluminum blades and the three-stage compressor. In addition, testing, maintenance, and refurbishment practices would be unchanged. Operational startup and shutdown resonance-stress-reduction practices had been developed over the years (since 1958, when the compressor first became operational). In addition, the rate at which the hub turned (i.e., the rpm) was varied during steady-state run to avoid resonance regions.

Alternative 2: New aluminum blades. In this alternative, worn-out blades would be replaced with new, high-aspect-ratio aluminum blades. Alternative 2 also included adding motorized stators in the compressor and revising internal guide vanes (IGVs). The project team assumed that startup and shutdown resonance stress reduction would be practiced as in Alternative 1, but that the motorized stators and IGVs would minimize the need for rpm changes during operation.

Alternative 3: Composite blades. This alternative included replacing current blades with new high-aspect-ratio composite blades, along with the added motorized stators and revised IGVs from Alternative 2. The engineers assumed that no resonance-avoidance activities or blade grinding would be required with this option.

After reviewing these alternatives, the PM conferred with the engineers. It came to light that the alternatives had been constrained by the collective intuition of the engineers about cost constraints. They did not want to propose alternatives that they thought would be too expensive. The PM, however, decided that if changes to the current compressor and motor were being considered anyway, there should be an alternative that included all the desired safety and operational changes. This brought the team to Alternative 4.

Alternative 4: Modified compressor. This alternative included replacing current blades with new high-aspect-ratio composite blades; purchasing motorized stators, revised IGVs, and new blade hubs; and making other modifications to enhance operation.

6.5 The Decision Model

The decision analyst treated this problem as a multiattribute decision problem involving cost and safety. To complete the decision analysis, he used the analytic hierarchy process (AHP) described in Section 4.8.

6.5.1 *The Hierarchy Diagram*

Figure 6-1 shows the hierarchy diagram for this decision analysis. The top level of the diagram contains the decision objective, which is phrased as best combination of LCC and safety. The middle level shows the two attributes against which each of the alternatives, on the third level, will be evaluated. After analyzing each alternative with respect to cost and safety, the analyst obtained normalized eigenvectors that rank each alternative with respect to cost and safety, respectively. These eigenvectors were combined to form a matrix. A normalized attribute vector represented the PM's belief about the relative importance of cost and safety. Combining both safety and cost, the dot product of the eigenvector matrix and the normalized attribute vector results in a vector that contains the relative rankings of each alternative, considering both attributes.

Although it is possible to use the AHP without quantitative analysis of cost and safety, doing so would simply reflect uninformed opinions. In this case, the PM wanted a quantitative analysis of the total LCC and a quantitative analysis of safety, so the analyst used the appropriate risk-assessment techniques.

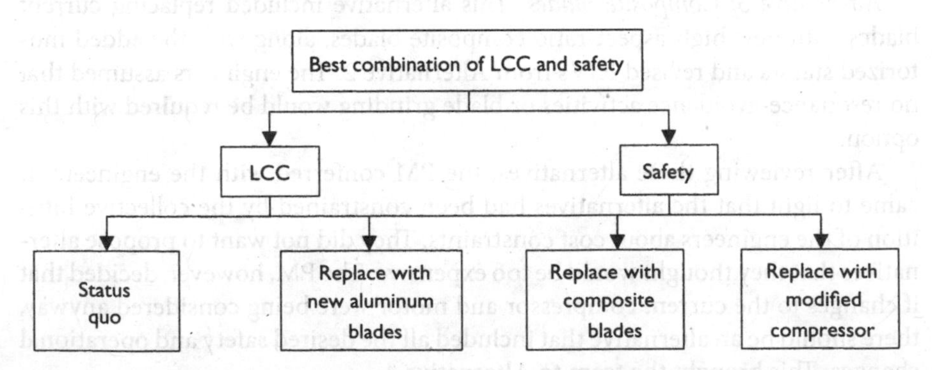

Figure 6-1. ***Hierarchy for Blade-Trade Study***

Note: LCC = life-cycle cost.

6.5.2 *Life-Cycle Cost Analysis*

LCC was defined as the amortized dollars per year of recurring and nonrecurring costs over the lifetime of the facility. Because we cannot know precise LCC in advance for postulated options, the analyst characterized the costs using probability distributions that expressed the uncertainty. The total calculated LCC included design, development, testing, analysis, hardware purchases, operation, maintenance, inspection, refurbishment, and cost of money. It also specifically included lost productivity costs associated with maintenance, inspections, blade change-outs, and resonance avoidance. Specifically, the cost model included the following:

- Material costs

— blades

— motorized stators

— IGVs

— hubs

— interstage stators

— clamshell modifications

- Materials used for research and development of alternatives
- Personnel costs for the following tasks

— normal operations

— blade and compressor inspections (every 25 hours of operation)

— blade grinding

— blade change-out

— other scheduled maintenance

— installation and assembly of new blades and compressor

— research and development of alternatives

- Energy cost of the 11-ft tunnel operation

All costs were calculated for 744 hours per year of test time for 30 years. The analyst accounted for the effect of resonance-avoidance procedures (i.e., lost productivity) by increasing the number of operating hours required to avoid resonance.

The uncertainties in material costs were derived from experience with previous major purchases at NASA. Personnel costs were based on the rate schedule of support contractors with uncertainty associated with the number of hours applied to each task. Maintenance cost uncertainties were taken from the variability of such costs in other facilities. Energy cost uncertainties were based on a statistical analysis of 20 months of energy usage and megawatt-hour prices for the UPWT. The analyst used the standard approach for combining probability distributions, Monte Carlo (or Latin Hypercube) simulation. Table 6-1 summarizes the mean results, and Figure 6-2 illustrates the uncertainty distributions.

Total costs were significantly lower for alternative 3 because the energy costs were lower. Energy costs were lower for composite blades because they did not require special operations to avoid damaging resonances while the compressor was increasing and decreasing in speed. Because of the dominance of energy costs, the relative ranking of alternatives was insensitive to large variations in costs for materials and labor.

Table 6-1. ***Mean Alternative Costs per Year Amortized over 30 Years***

Alternative	*Material costs*	*Operating staff costs*	*Maintenance, inspections, and staff costs for development*	*Energy costs*	*Total*
1. Status quo	2.9E+05	4.0E+05	2.3E+05	3.2E+06	4.1E+06
2. Aluminum blades	5.0E+05	4.0E+05	2.4E+05	2.8E+06	3.9E+06
3. Composite blades	8.5E+05	4.0E+05	1.3E+05	2.0E+06	3.4E+06
4. Modified compressor	1.5E+06	4.0E+05	1.9E+05	2.0E+06	4.1E+06

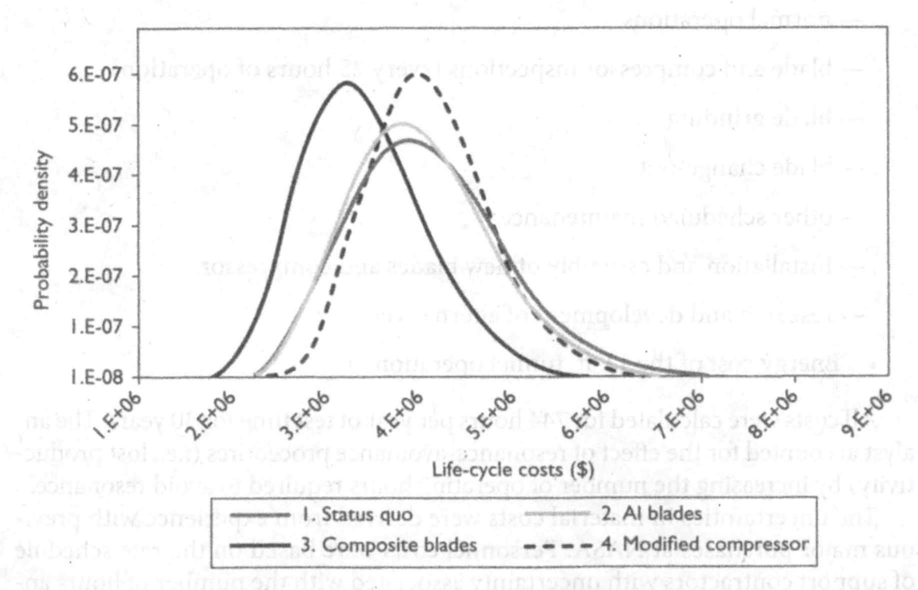

Figure 6-2. ***Total Life-Cycle Costs (LCC) by Alternative***

6.5.3 *Safety Analysis: Catastrophic Wind Tunnel Damage from Compressor Blades*

Both human error (e.g., foreign objects inadvertently left in the tunnel or mistakes in setting up the model) and "spontaneous" failures (e.g., overstress) contribute to catastrophic damage from compressor blades. Design flaws and equipment failures cause spontaneous failures, which can occur either early in operational life or after many years of operation. Early incidents tend to be caused by (1) operational characteristics that were unknown, (2) design mistakes, or (3) inadequate quality control or configuration management. Incidents that occur later in operational life tend to be caused by worn-out components. Events induced by foreign objects appear to occur at any time during the tunnel's lifetime. Because the forecasting of future failures relies on history, it is inherently uncertain. In this case, the analyst used the past catastrophic failure history of other tunnels— combined with that of the UPWT—as a basis for forecasting the probable range of future catastrophic failure frequencies. Bayesian analysis (see Section 3.6.1) was used to combine the history of other tunnels with that of the UPWT to produce the forecasted result.

The decision analyst took the following steps to estimate the frequency of catastrophic failure:

- Compiled catastrophic failure incidents from wind tunnels around the country. Data were collected in the two general causal categories (human error and spontaneous failure). Table 6-2 summarizes the resulting data.

- Developed probability distributions over the range of frequencies that represent the compiled history. The analyst developed two such distributions. The first represented the uncertainty in the frequency of catastrophic damage resulting from spontaneous failures, and the second represented the uncertainty in the frequency of catastrophic damage caused by foreign objects.

- Summed the probability distributions to arrive at the total estimated compressor blade related risk in terms of the incidents of catastrophic damage per year.

The analyst did not distinguish among different levels of catastrophic failure because the PM thought this was unnecessary. Unfortunately, the compiled records were not explicit enough to suggest trends between metal and composite blades.

The analyst performed two Bayesian analyses to obtain the estimated risk of catastrophic failure. The analyst (1) developed a prior probability distribution using current knowledge and the data from wind tunnels other than the UPWT and (2) modified this distribution (refer to Equation 3-1) by using data specific to the UPWT.

Foreign-Object-Caused Damage Risk Assessment. From Table 6-2, an estimate of the mean foreign-object-caused catastrophic failure rate for wind tunnels, excluding the UPWT, equals 10/(49)(28) = 0.007 per tunnel-year. A risk assessment, though, requires an uncertainty, which can be obtained by incorporating the underlying failure mechanism (human errors, in this case). Generic human error rates for routine maintenance operation actions, which have well-rehearsed procedures with no significant time constraints, tend to have a range of failure probability per action of 0.01 to 10^{-4}. The analyst took these limits to be a 90% uncertainty interval. Based on a human reliability analysis model (e.g., Swain and Guttman 1982) the distribution was assumed to be lognormal. Equation 6-1 shows the lognormal probability density function.

$$l(x) = \frac{1}{\sigma\sqrt{2\pi}\,x} e^{-\frac{1}{2}\left(\frac{\ln x - u}{\sigma}\right)^2} \tag{6-1}$$

In Equation 6-1, μ and σ are parameters with the following properties:

- the mean of the lognormal distribution = $e^{\mu + \frac{1}{2}\sigma^2}$ and,
- the median of the lognormal distribution = e^{μ}.

The analyst sought a distribution for the prior (see Equation 3-1) that could represent both the estimated mean from the compiled data and the estimated uncertainty derived from human error rates. They developed a lognormal distribution with the parameters $\mu = -5.94$ and $\sigma = 1.4$ as the prior.

A Poisson distribution is an appropriate likelihood function for failure rates over time. Equation 6-2 shows the appropriate form of the Poisson distribution for this case study.

$$p(k,t \mid \lambda) = e^{-\lambda t} \frac{(\lambda t)^k}{k!} \tag{6-2}$$

Table 6-2. *Wind Tunnel Catastrophic Failure Survey*

Information Collected	*Data*
Number of wind tunnels surveyed	50
Number of spontaneous catastrophic failures other than UPWT	16
Number of spontaneous failures in UPWT	1
Number of foreign-object-caused catastrophic failures other than UPWT	10
Number of foreign-object-caused catastrophic failures in UPWT	0
Number of runs per year per tunnel	25 to 200
Average operating years per tunnel other than UPWT	28
Operating years of UPWT	36

Note: UPWT = Unitary Plan Wind Tunnel.

In Equation 6-2, $p(k, t \mid \lambda)$ represents the probability of observing k failures in time t given the prior failure rate is λ.

Because the analyst constructed the prior using general knowledge and data from wind tunnels other than the UPWT, it was appropriate to include the UPWT data of 0 catastrophic foreign-object-caused damage incidents in 36 years (from Table 6-2) as k and t, respectively. The resultant prior and updated distributions are shown in Figure 6-3. Because the UPWT record was better than the cumulative record of other wind tunnels, the updated distribution shifts to the left toward lower frequency. The mean of the updated distribution is 0.004 per tunnel-year.

Spontaneous Damage Risk Assessment. The data in Table 6-2, disregarding the UPWT, shows 16 failures over 49 tunnels with an average of 28 years of operation per tunnel. An estimate of the mean frequency of the prior distribution for spontaneous catastrophic damage would be, therefore, 0.012 per tunnel-year. No tunnels in the survey had more than 1 catastrophic failure in 29 years, which indicates that a reasonable upper 95th percentile would be 0.03/tunnel-year. To create a reasonable prior in the blade-trade case study, the analyst used a Beta distribution (Equation 6-3) with the 0.012 per tunnel-year mean and the 95th percentile as follows:

$$b(x) = \frac{x^{\alpha-1}(1-x)^{\beta-1}}{B(\alpha,\beta)} \tag{6-3}$$

where $B(\alpha,\beta)$ is a Beta function, $\alpha = 1.6$, and $\beta = 129$.

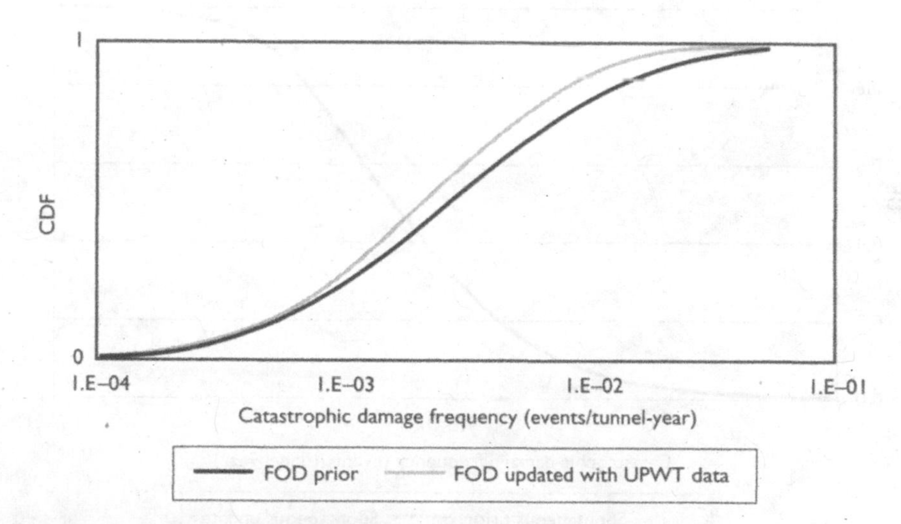

Figure 6-3. ***Prior and Updated Probability Distribution for the Frequency of Catastrophic Foreign Object Damage***

Notes: At the Unitary Plan Wind Tunnel (UPWT). CDF = cumulative distribution function; FOD = foreign object damage.

To develop the updated frequency of spontaneous catastrophic failure, the analyst combined Equations 6-2 and 6-3 in accordance with Equation 3-1. Because the prior was constructed using general knowledge and data from wind tunnels other than the UPWT, it was appropriate to include the UPWT data of 1 catastrophic foreign-object-caused damage incident in 36 years (from Table 6-2) as k and t, respectively, as shown in Equation 6-2. Figure 6-4 shows the resultant prior and updated distributions. Because the UPWT record was worse than the cumulative record of other wind tunnels, the updated distribution in the figure shifts to the right (toward higher frequency). In this study, the mean of the updated distribution was 0.016 per tunnel-year.

Combining the Two Blade Damage Frequencies. Using a standard Monte Carlo technique, the analyst obtained the total blade damage distribution by summing the two updated distributions shown in Figures 6-3 and 6-4. Figure 6-5 illustrates the resultant total estimated catastrophic damage probability distribution caused by blade damage. Because there were no recorded data that would allow an analysis of actual damage caused by aluminum blades and composite blades, this part of the risk assessment could not be done quantitatively. The project engineers and managers, however, strongly preferred composite blades; aluminum blades can cause a great deal of damage because they don't break up on impact. If an aluminum blade hits an object edge-on, it acts like a knife cutting through the affected equipment. The composite blades, on the other hand, tend to shatter on impact, causing much less equipment damage.

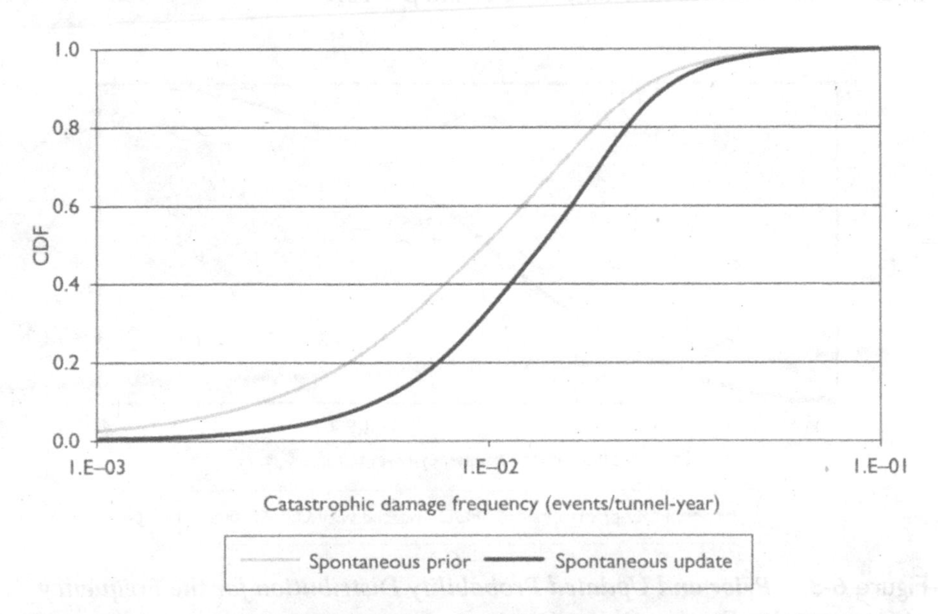

Figure 6-4. ***Prior and Updated Probability Distribution for the Frequency of Catastrophic Spontaneous Failures***

Notes: At the Unitary Plan Wind Tunnel (UPWT). CDF = cumulative distribution function.

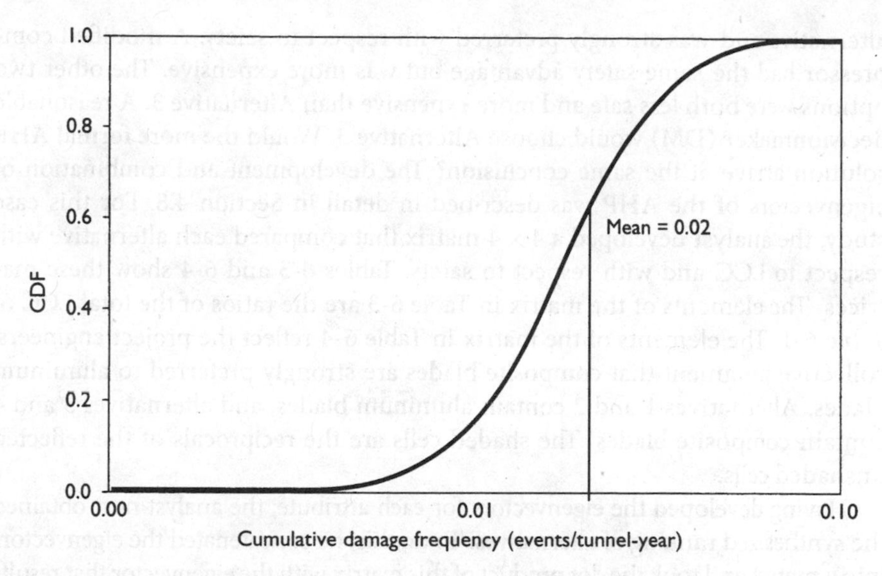

Figure 6-5. ***Total Estimated Frequency of Catastrophic Failures Caused by Blade Damage***

Notes: At the Unitary Plan Wind Tunnel (UPWT). CDF = cumulative distribution function.

6.6 The Value Model

As described in Section 4.8, the AHP has three steps: (1) determine the importance of each option (at the third level) with respect to each criterion (at the second level); (2) determine the importance of each criterion (second level) with respect to the goal (first level); and (3) combine the first two steps to obtain the rank of each option (third level) with respect to the goal (first level).

The analyst's next task was to define the matrix that ranks the importance of LCC against safety with respect to the goal in the top box of Figure 6-1 (best combination of LCC and safety). If the PM had been able to communicate the relative importance of safety versus LCC in a straightforward way, the AHP eigenvector solution would have produced a specific ranking of alternatives. I presented a useful alternative in Section 4.8.2, in which this judgment is left as a sensitivity parameter and a decision trajectory of each alternative is produced.

6.7 The Rankings, Obtained by Synthesizing the Decision and Value Models

The proper decision in this study was made reasonably clear by the analysis. Acquiring new compressor blades (Alternative 3), was clearly the least-cost alternative and was strongly preferred with respect to safety. A modified compressor had the same safety advantage but was more expensive. The other two options were both less safe and more expensive than Alternative 3. A reasonable decisionmaker (DM) would choose Alternative 3. Would the more formal AHP solution arrive at the same conclusion? The development and combination of eigenvectors of the AHP was described in detail in Section 4.8. For this case study, the analyst developed a 4 × 4 matrix that compared each alternative with respect to LCC and with respect to safety. Tables 6-3 and 6-4 show these matrices. The elements of the matrix in Table 6-3 are the ratios of the total LCC of Table 6-1. The elements of the matrix in Table 6-4 reflect the project engineers' collective judgment that composite blades are strongly preferred to aluminum blades. Alternatives 1 and 2 contain aluminum blades, and alternatives 3 and 4 contain composite blades. The shaded cells are the reciprocals of the reflected unshaded cells.

Having developed the eigenvectors for each attribute, the analyst next obtained the synthesized ranking of alternatives. To do this, he concatenated the eigenvectors into a matrix and took the dot product of this matrix with the eigenvector that results from the importance ratio of safety to cost. Using the concept of decision trajectories, the analyst evaluated the eigenvector of the matrix in Table 4-5 multiple times, each time using a different number for w_s. Recall from Table 4-11 that each value of the relative importance of safety to cost, η, provides a unique safety weighting factor, w_s. The dot product was performed in order to obtain a point on the decision trajectory. Figure 6-6 shows the resultant decision trajectories for the alternatives in this study.

The results make intuitive sense. As the importance of safety over cost increased, the two alternatives with the advantage in safety became increasingly preferred over the other two options. At the left side of the figure, for w_s approaching 0, the relative ranks approach those of the LCC eigenvector. On the opposite side, as w_s approaches 1, the relative ranks approach those of the safety eigenvector. We can see that Alternative 3 is always superior to the others, but it converges with Alternative 4 as LCC is reduced in importance (i.e., when w_s approaches 1) because they have identical safety preferences. Similarly, Alternatives 1 and 2 converge as w_s approaches 1.

Table 6-3. *Life-Cycle Cost (LCC) Matrix*

	1. Status quo	*2. Aluminum blades*	*3. Composite blades*	*4. Modified compressor*
1. Status quo	1	0.95	0.83	1
2. Aluminum blades		1	0.87	1.05
3. Composite blades			1	1.2
4. Modified compressor				1

Note: The eigenvector of the principal eigenvalue of this matrix is (0.235, 0.247, 0.283, 0.235).

Table 6-4. *Safety Matrix*

	1. Status quo	*2. Aluminum blades*	*3. Composite blades*	*4. Modified compressor*
1. Status quo	1	1	0.2	0.2
2. Aluminum blades		1	0.2	0.2
3. Composite blades			1	1
4. Modified compressor				1

Note: The eigenvector of the principal eigenvalue of this matrix is (0.083, 0.083, 0.417, 0.417).

6.8 The Outcome

What really happened in the UPWT case study? The PM chose Alternative 3, and composite blades were purchased and installed.

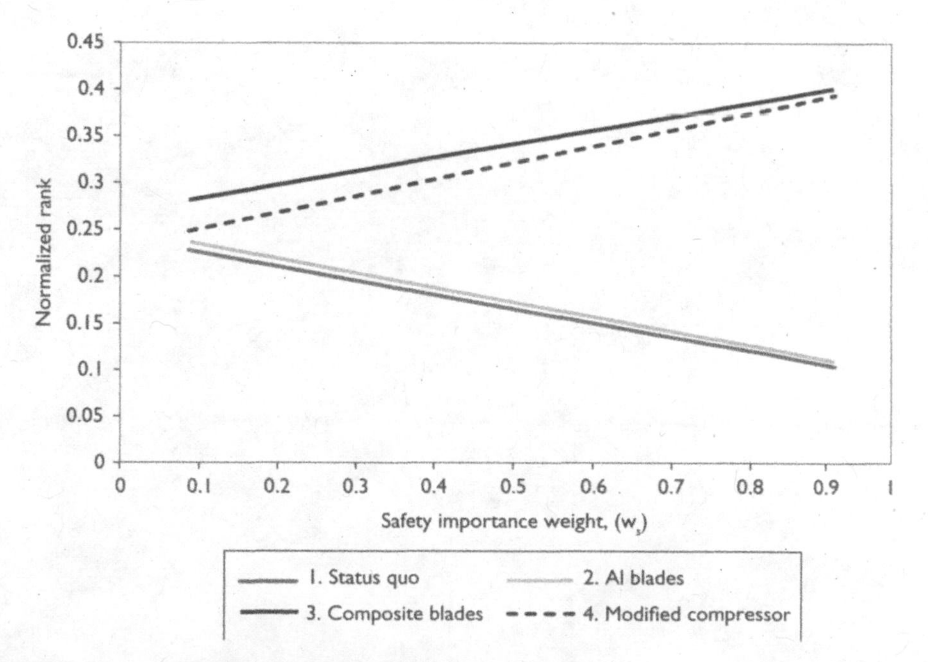

Figure 6-6. *Blade-Trade Study Decision Trajectories*

Note

1. In this chapter, I refer to the 11 × 11 ft transonic wind tunnel.

Reference

Swain, A., and H. Guttman. 1982. *Handbook of Human Reliability Analysis with Emphasis on Nuclear Power Plant Applications.* NUREG/CR-1278. Washington, DC: U.S. Nuclear Regulatory Commission.

CHAPTER 7

Choosing among Space Shuttle Auxiliary Power Unit (APU) Safety Improvement Strategies

AFTER THE STS 51L ACCIDENT in 1986, NASA commissioned a study to investigate the efficacy of probabilistic risk assessment (PRA) for future use to help with safety improvements on the space shuttle. Modern PRA had not previously been used at NASA. The study, which was performed on the shuttle orbiter's APUs, resulted in a probabilistic risk assessment proof-of-concept (POC) study (MDAC 1987). In this chapter I describe an example decision analysis, following the POC, requested by the associated administrator for safety, which demonstrated how a PRA can be used to choose among safety improvement strategies. It was a method demonstration study about different decision models that can be applied using the results of a PRA. No action was taken on the results.

7.1 How the APUs Work

As depicted in Figure 7-1, three APUs reside in the aft compartment of the shuttle's orbiter, aft of the 1307 bulkhead. Each APU is the only source of power to a dedicated hydraulic pump. The hydraulic pumps on the APUs are necessary to pressurize a connected hydraulic system that powers the following mission-critical functions:

- Throttling the main engines during ascent
- Controlling the thrust vector control powering the gimbals of the main engine nozzles during ascent
- Retracting the external tank umbilical plates during ascent
- Controlling the rudder and other aerosurfaces during descent

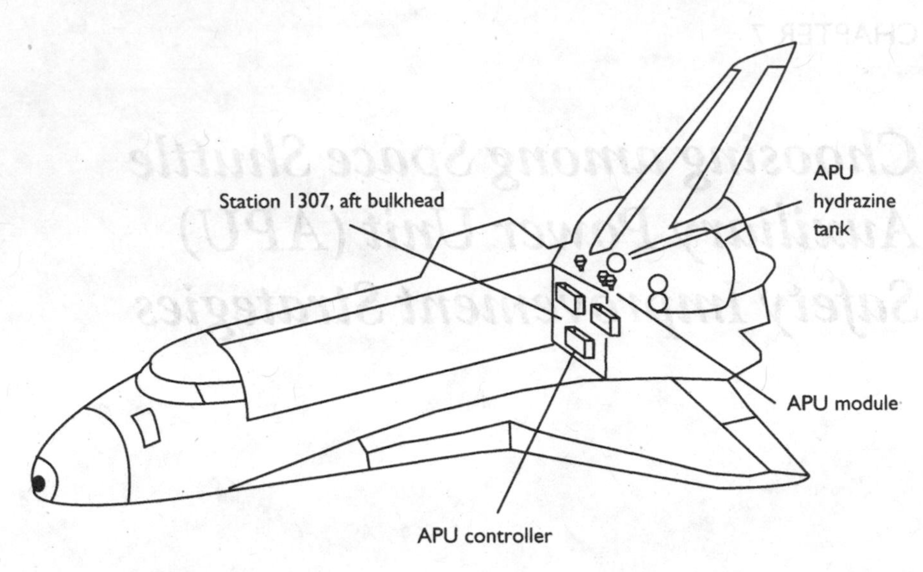

Figure 7-1. *Locations of Three Auxiliary Power Units (APUs)*

- Deploying landing gear
- Braking on the runway after landing and
- Differential steering on the runway after landing

Figure 7-2 shows a schematic of an APU, which powers the pump through a direct shaft/gear linkage. The shaft is rotated by a turbine that normally spins at 74,000 rpm. A catalytic decomposition of liquid hydrazine (N_2H_4) produces a high-temperature (about 817°C [1,500° F]) expansion of gas, which pushes the turbine. The turbine and gear system requires lubricating oil that is run through a cooled lube oil subsystem. The lube oil pump is also powered by the turbine. The turbine exhaust flow returns over the exterior of the gas generator, cooling it, and is then directed overboard through an exhaust duct at the upper portion of the aft fuselage near the vertical tail.

7.2 What Can Go Wrong

Leaking hydrazine causes damage stemming from three of its physical properties. First, hydrazine is corrosive to certain materials. One such material is the Kapton wiring insulation used extensively in the aft compartment. Second, hydrazine is flammable in as little as about 4.7% oxygen. Hot spots on the APUs themselves can provide an ignition source for a hydrazine–oxygen mixture. Third, hydrazine will autodecompose to nitrogen, hydrogen, and ammonia in an exothermic reaction when it comes in contact with certain materials, such as metal oxides.

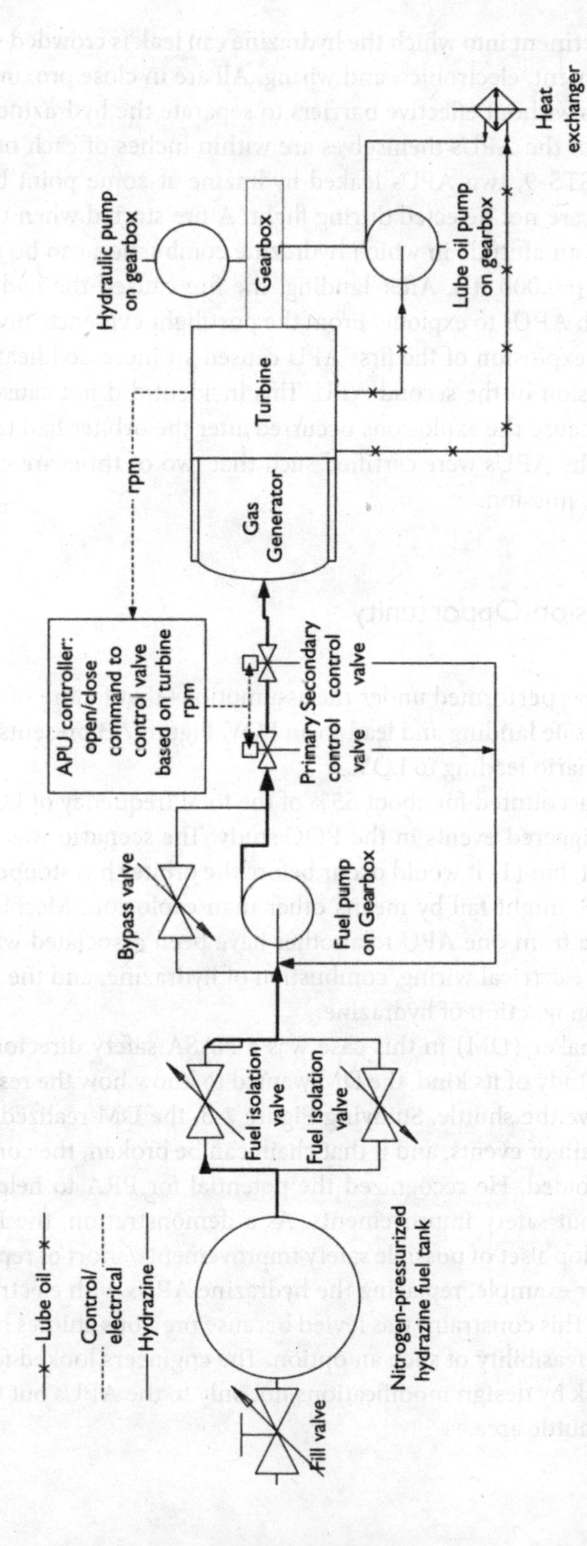

Figure 7-2. ***Auxiliary Power Unit (APU) Schematic***

The aft compartment into which the hydrazine can leak is crowded with main propulsion equipment, electronics, and wiring. All are in close proximity to the hydrazine sources without effective barriers to separate the hydrazine from the equipment. Two of the APUs themselves are within inches of each other. During the flight of STS-9, two APUs leaked hydrazine at some point before descent. The leaks were not detected during flight. A fire started when the orbiter had descended to an altitude in which hydrazine combustion can be supported (about 18,290 m [60,000 ft]). After landing, the fire caused the hydrazine retained within both APUs to explode. From the postflight evidence, investigators deduced that the explosion of the first APU caused an increased heat load and subsequent explosion of the second APU. This incident did not cause a loss of vehicle (LOV) because the explosions occurred after the orbiter had landed and come to a stop. The APUs were certified such that two of three are considered required for a safe mission.

7.3 The Decision Opportunity

The POC study was performed under the assumptions that failure of two APUs would preclude a safe landing and lead to an LOV. Figure 7-3 presents the highest frequency scenario leading to LOV.

This scenario accounted for about 65% of the total frequency of LOV resulting from APU-triggered events in the POC study. The scenario was similar to the STS 9 incident, but (1) it would occur before the orbiter has stopped moving and (2) two APUs might fail by means other than explosion. Mechanisms for spreading damage from one APU to another have been associated with hydrazine contact with electrical wiring, combustion of hydrazine, and the spontaneous decomposition reaction of hydrazine.

The decisionmaker (DM) in this case was a NASA safety director. Because this was the first study of its kind, the DM wanted to know how the results could be used to improve the shuttle. Studying Figure 7-3, the DM realized that such a scenario is a chain of events, and if that chain can be broken, the consequence (LOV) can be avoided. He recognized the potential for PRA to help with decisionmaking about safety improvements. As a demonstration, the DM asked engineers to develop a set of possible safety improvements, short of replacing the APUs entirely. For example, replacing the hydrazine APUs with electrical APUs was out of scope. This constraint was levied because previous studies had shown the low technical feasibility of such an option. The engineers looked for ways to improve safety risk by design modifications not only to the APUs but to the surrounding space shuttle area.

7.4 The Problem Statement

The DM gave the engineers and the decision analyst (the project team) a problem statement paraphrased as follows: "Show me ways to improve the APUs, using PRA, but being mindful of cost. Show me how to prioritize the suggested improvements in a quantitative way that mitigates the effects of bias, opinion, judgment, and assumption. Furthermore, don't show me any ideas that cannot be feasibly installed and operated on the space shuttle."

7.5 The Objective and the Attributes

The DM's problem statement was explicit enough to allow the project team to quickly formulate an objective and attributes. The objective was stated as follows: Find the best APU safety improvement strategy for the hydrazine leakage scenario, also considering cost as an attribute. Technical feasibility (i.e., ability to develop, install, and operate the modification) was considered a constraint. The team would not carry any technically infeasible alternatives through the analysis or present them to the DM. The attributes, therefore, were limited to (1) safety, as measured by LOV frequency resulting from APU-initiated problems (not just leaks), and (2) total development and installation cost of the modification.

7.6 The Alternatives

After a number of feasibility studies, the engineers proposed six safety improvement strategies. I give the details for each prospective strategy in the sections that follow.

Strategy A: Improve leak detection system. At the time of this study, the APU hydrazine fuel lines were instrumented with a number of temperature sensors. These were installed to alert the crew, via the caution and warning system, of temperatures approaching hydrazine's freezing point on the low end and its boiling point on the high end. The alarms were set at approximately 7°C (45°F; the low end) and 93°C (200°F; the high end). These alerts were used in orbit only. There were no hydrazine fuel temperature alerts during launch, ascent, entry, descent, or landing. Hydrazine pressure was monitored only at the fuel tank, and an alert was given if the pressure exceeded 2413 kPa (350 psi) or dropped below 758 kPa (110 psi).

Because no previous leaks, including those on STS 9, had been detected, this instrumentation was clearly deficient for leak detection. The project team decided that a leak detection and warning system should consist of pressure sensors, transmitters, controllers, wires, and other hardware and software needed to

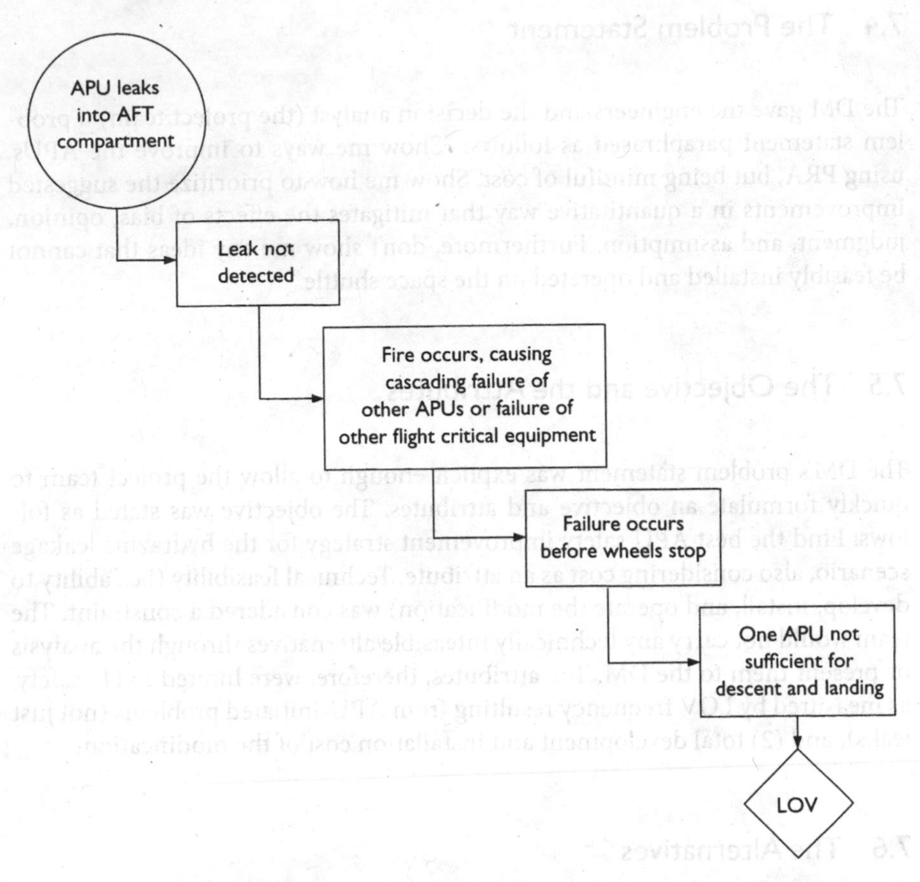

Figure 7-3. ***Hydrazine Leak Scenario Could Lead to Space Shuttle Loss of Vehicle (LOV)***

sense and transmit signals to the general-purpose computers (GPCs) for crew warning.

Strategy B: Provide barriers. This strategy was conceived to address the failure of two APUs by fire or explosion during descent or landing. Two of the APUs (APU 1 and APU 2) were within a few inches of each other on the port side of the aft compartment of the orbiter. The third APU (APU 3) was a few feet away on the starboard side. APU 3 was in relatively close proximity to avionics bays that house flight-critical components—failures of these components could lead to LOV. Strategy B involved barriers separating APU 1 from APU 2 and separating APU 3 from the avionics bays. The barriers need only be effective against fire and explosion during the last 10 minutes of the flight because the atmosphere cannot support hydrazine combustion until an altitude of about 18,290 m (60,000 ft) is reached. The design objectives for these barriers were to

- Delay a fire or explosion occurring in APU 1 or APU 2 from spreading to the adjacent APU before the orbiter has stopped,
- Delay a fire or explosion occurring in APU 3 from spreading to adjacent avionics bays, or
- Delay the spread of hydrazine (which will attack the insulation material) from an APU to an adjacent APU or flight-critical equipment.

The requisite barriers could be created by surrounding each APU unit with a "can." The can material could be a high-temperature metal, such as titanium; a strong, high-temperature ceramic; or a composite.

Strategy C: Achieve inert atmosphere in aft compartment. Fires were extremely unlikely during launch because the aft compartment was pressurized with nitrogen before launch. This kept the ambient oxygen content of the aft compartment below the 4.7% needed for hydrazine combustion until altitude provided a sufficiently oxygen-free environment. Engineers suggested a system that could make the environment around the APUs inert, which would prevent combustion during descent and landing. The system would not need to be activated until the orbiter descended to an altitude of about 18,290 m (60,000 ft) during descent. An aft compartment "inerting" system would consist of tanks containing pressurized nitrogen, gas dispersion units, isolation valves, associated piping, controllers and switches for crew actuation, associated circuitry connected to the GPCs for crew alarm, software for crew warning, ground telemetry, and a system actuation device.

Strategy D: Achieve inert atmosphere within barriers only. This option was a combination of Strategies B and C. Each barrier can would be connected to a supply of inert gas as in Strategy C. There would be no need to inert the entire aft compartment.

Strategy E: Detect and suppress fire in the environment around the APUs. This concept would be most efficient if the detectors were located within the barrier cans of Strategy B. Under Strategy E, engineers would design and install smoke, particle, and temperature detectors with signal transmitters within each barrier; fire-suppression equipment connected to each barrier; and associated hardware and software for crew alarm with the ability of manual and automatic response.

Strategy F: Change APU/hydraulics design. With this strategy, the engineers aimed to make one of three APUs (instead of two of three) sufficient for safe operation. The flight-control surfaces (the elevons, the rudder/speedbrake, and the body flap) are driven by hydraulic actuators. These are, in turn, enabled by a hydraulic pump that pressurizes the hydraulic system. Each APU powers a hydraulic pump. If that pump could not supply sufficient pressure, it would be isolated from the corresponding hydraulic actuators and another APU/hydraulic system would be selected to power the flight-control surfaces. In the existing design, two of three APU/hydraulic pumps were required for 100% capability of the actuators. If only one of three APUs were inoperable, the flight-control surfaces would operate at approximately one-half speed. The engineers proposed modifying the APU/hydraulics such that any single APU/hydraulic system is capable of powering a 100% capability hydraulic motor. This would have required modification of the hydraulic pump and system to operate at a higher pressure, along with an increase in APU power.

Status quo: Do nothing. In a decision analysis, it is always good practice to include continuation of current practice as a basis for comparison.

7.7 The Decision Model

The decision analyst treated this demonstration study as a multiattribute decision problem involving cost and safety, so he decided to compare several different decision models and rules:

- Maximize the benefit-to-cost ratio in which benefit would be measured in reduction of LOV frequency associated with each modification and cost would be the total cost for development and installation.
- Minimize the expected impact (often called expected loss) so that the expected impact would be the sum of the total development and installation cost and the expected replacement cost.
- Create a decision tree, in which the expected cost would be maximized and safety would be converted to a cost metric.
- Develop a decision tree that would maximize the weighted safety and cost utility of a risk-neutral DM (see Equation 4-12).
- Conduct an analytic hierarchy process (AHP) selecting the highest combined cost and safety ranking without using utility functions to establish preferences.
- Conduct an AHP selecting the highest combined cost and safety ranking using utility functions of a risk-neutral DM to establish preferences.
- Pay attention to intuition.

As a first step, the decision analyst developed a qualitative decision tree, which is shown in Figure 7-4. This diagrams the six strategies and the chance nodes. The decision tree is complete when both the safety and costs have been calculated.

7.7.1 Safety Risk Assessment

Safety was measured as LOV frequency (per mission) resulting from APU-triggered events. The POC study provided the baseline LOV frequency against which the decision analyst compared each strategy, as shown in the column of

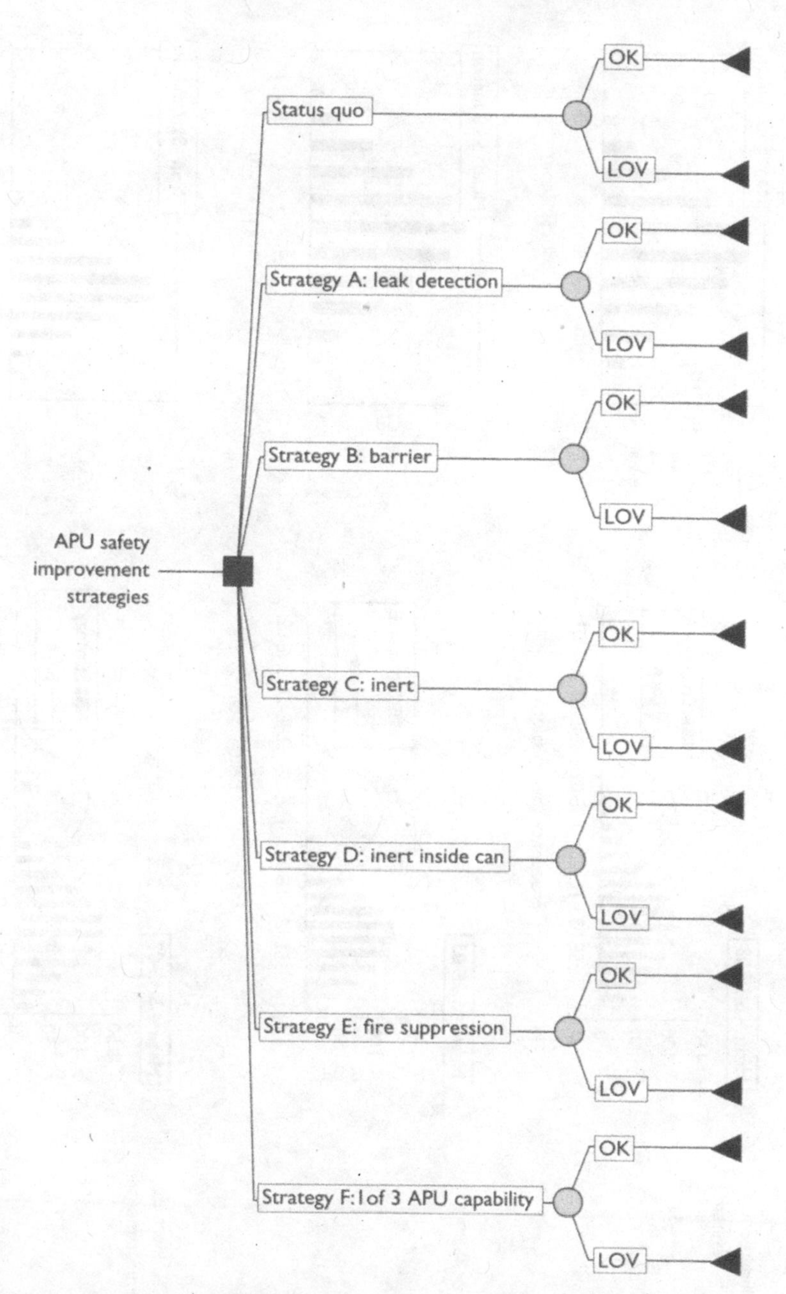

Figure 7-4. ***Qualitative Decision Tree for Auxiliary Power Unit (APU) Safety Improvement Decision Analysis***

Note: LOV = loss of vehicle.

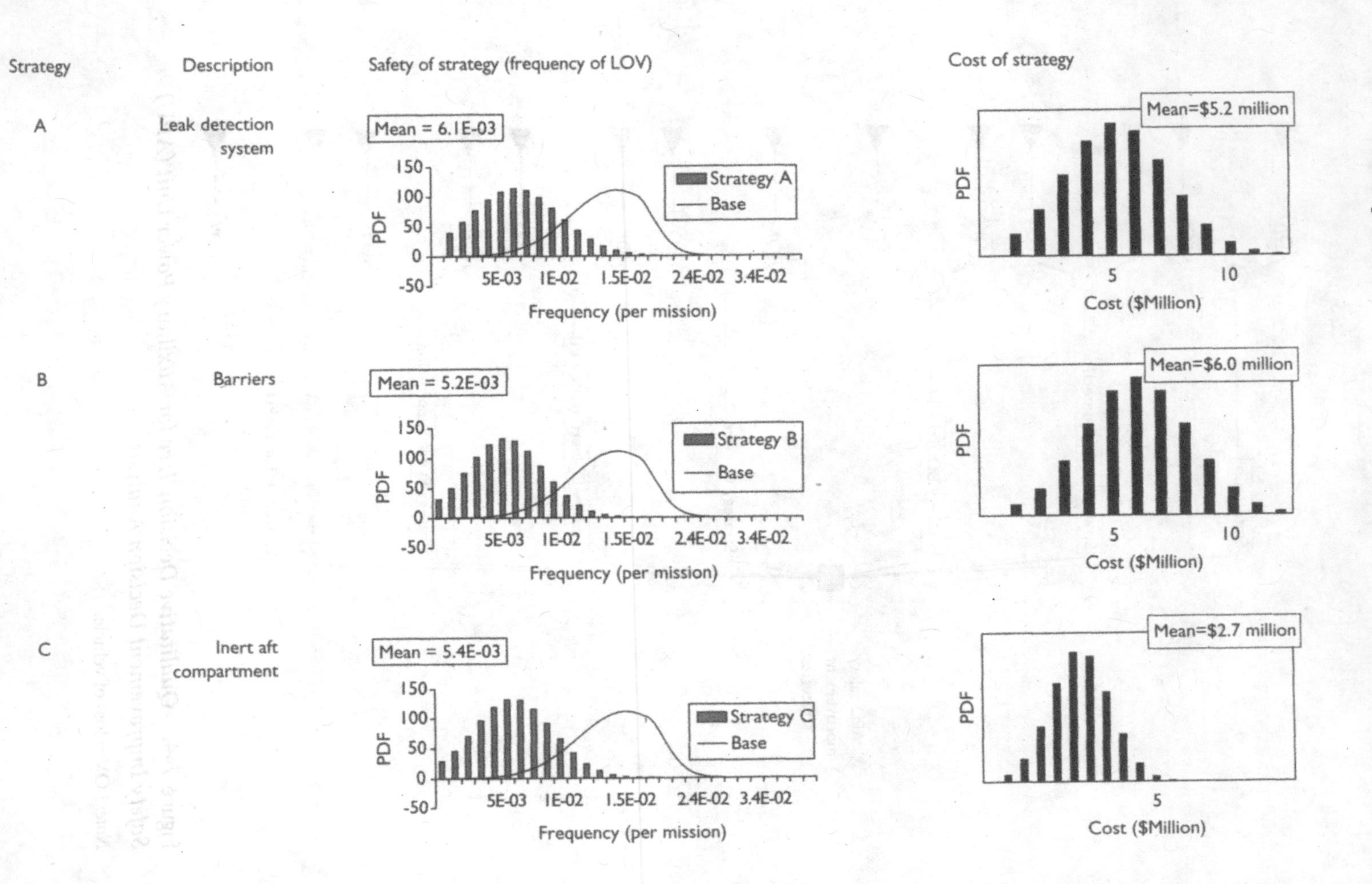
Strategy
Description
Safety of strategy (frequency of LOV)
Cost of strategy
A
Leak detection system
Mean = 6.1E-03
Strategy A
Base
PDF
150
100
50
0
-50
5E-03
1E-02
1.5E-02
2.4E-02
3.4E-02
Frequency (per mission)
Mean=$5.2 million
5
10
Cost ($Million)
B
Barriers
Mean = 5.2E-03
Strategy B
Base
Mean=$6.0 million
C
Inert aft compartment
Mean = 5.4E-03
Strategy C
Base
Mean=$2.7 million

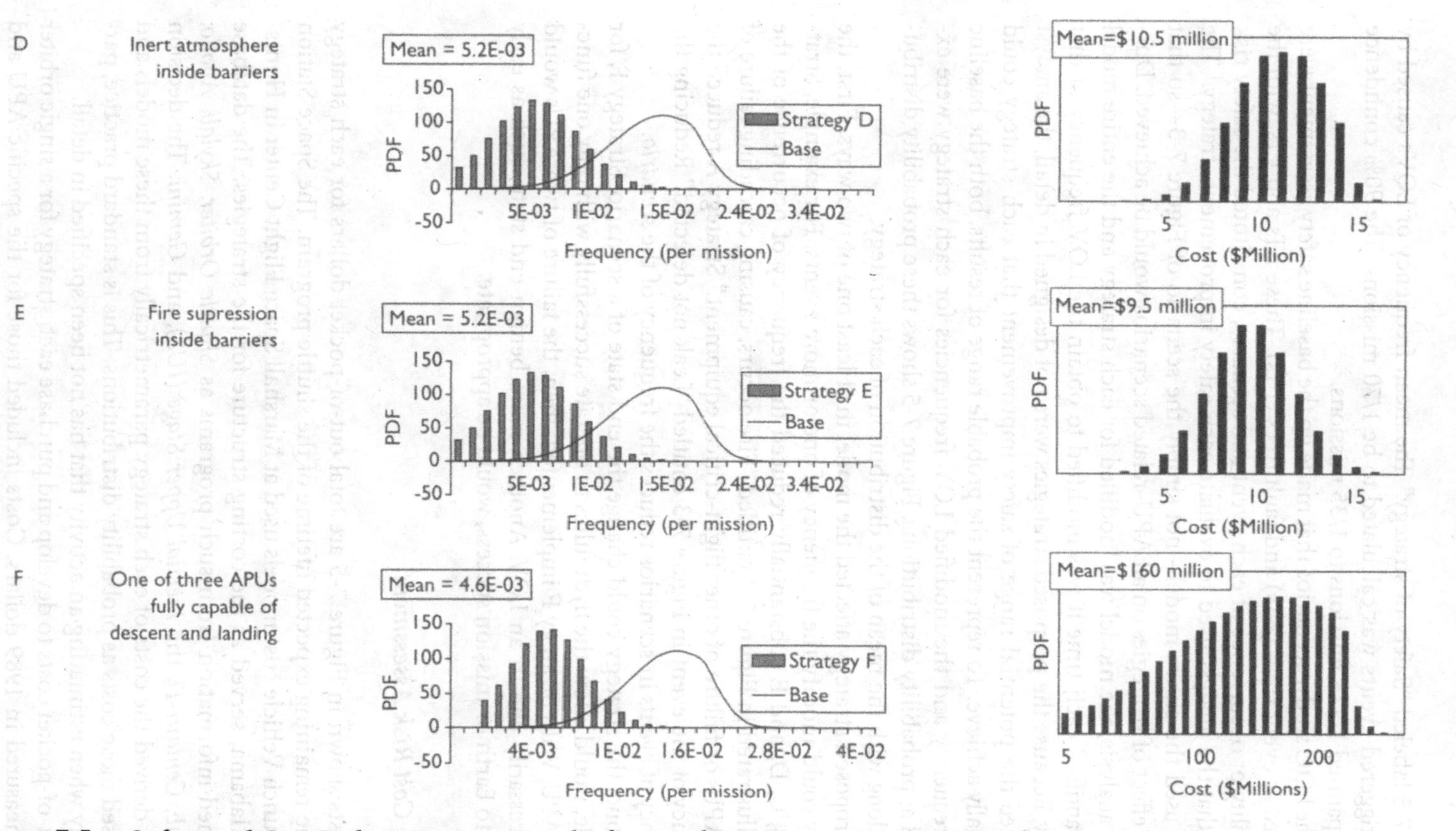

Figure 7-5. ***Safety and Cost Risk Assessment Results for Auxiliary Power Unit (APU) Safety Improvement Decision***

Notes: LOV = loss of vehicle; PDF = probability density function.

Figure 7-5 labeled "Safety of strategy." The mean frequency of LOVs caused by APU-triggered events was calculated to be 1/70 missions. The 90% confidence range spanned 1/215 missions to 1/35 missions.

In the POC study, scenarios that made up the baseline safety assessment were exhibited as event trees (ETs) and fault trees (FTs). These ETs and FTs, with the probability of occurrence of each event and scenario, constituted the safety risk model that the analyst used to evaluate each safety improvement strategy.[1] The analyst used the entire model—not simply the scenario of Figure 7-3—so that the net effect of strategies on all APU-related scenarios would be achieved. During the analysis, the model was modified for each strategy and the entire model was quantified each time it was modified to obtain the LOV frequency of that strategy. Because the proposed strategies were not designed in detail, the analyst estimated the potential range of safety improvement that each strategy could potentially achieve. To represent the probable range of results, both the baseline LOV frequency and the modified LOV frequencies for each strategy were expressed as probability distributions. Figure 7-5 shows these probability distributions, along with the mean of the distribution of each strategy.

A proposed strategy affected the model in at least one of two ways. First, the strategy could modify the frequency of one or more events. For example, Strategies B, C, D, and E substantially reduced the frequency of occurrence of the event illustrated in Figure 7-3 entitled "Fire occurs, causing cascading failure of other APUs or failure of other flight-critical equipment." Strategy A reduced the frequency of the event in Figure 7-3 entitled "Leak not detected." Reducing the frequency of *events* in scenarios reduces the frequency of the *scenario*.

Second, the strategy could change the end state of a scenario. Strategy F, for example, would allow the hydraulics to operate successfully with only one functional APU. With Strategy F implemented, then, the failure of two APUs would not necessarily end in an LOV. Another, more benign end state, such as early return to Earth or mission success, would be appropriate.

7.7.2 Cost Risk Assessment

The costs shown in Figure 7-5 are total out-of-pocket dollars for each strategy over the remaining expected lifetime of the shuttle program. The Space Station and Launch Vehicle cost models used at Marshall Space Flight Center in Huntsville, Alabama, served as the costing structure for the strategies. The database contained information from such programs as *Shuttle Orbiter, Skylab, Apollo, Spacelab, Centaur-D,* the *Inertial Upper Stage (IUS),* and *Gemini.* The decision analyst derived the costs of each strategy parametrically from these models and expressed those costs as probability distributions. This is standard practice, particularly when estimating an activity that has not been specified in detail.

Out-of-pocket costs to develop and purchase each strategy for a single orbiter were measured in 1989 dollars. Costs included those for the specific APU and aft compartment changes, as well as changes to structure, thermal systems, the electrical power system, the hydraulic system, and ground and flight software. Other costs were for the test articles, ground support equipment, system engineering, and integration and management functions normally associated with retrofitting an orbiter. For this study, the analyst took the baseline out-of-pocket costs, which applies to maintaining the status quo, as zero dollars. The costs shown in Figure 7-5 represent the recurring and nonrecurring costs of a single orbiter except that operational spares and fleet retrofit costs were excluded. Estimated changes in subsystem mass were an integral part of the cost evaluation.

7.7.3 *Intuitive Evaluation of Risk Results*

Figure 7-5 shows the safety curve of each strategy compared to the baseline distribution, making the net safety improvement evident. Although the curve of each strategy is overlapped by the baseline curve, we can see that each strategy reduces the frequency of LOV associated with APU-triggered events. The net safety benefit of a strategy can be obtained by subtracting the baseline curve from the strategy curve. The similarity of the safety curves viewed by themselves does not give a DM much guidance about which strategy to choose. In practical terms, the safety improvement benefits of the strategies are quite similar. We can also see that four of the six cost curves are quite similar. Strategy F, however, is significantly more costly than the others—even its lower bound is larger than the upper bound of the other options. Strategy C is clearly less costly than the other options. Its mean is significantly lower and the curve is narrower, signifying more confidence in the cost estimates. Simply on the basis of what appears to be straightforward intuitive reasoning about the curves, considering only risk and cost as relevant attributes, Strategy C would be preferred over the others. But even with two attributes, differing perspectives and decision rules will offer different insights into the decision and, as we'll see, can lead to different recommended strategies.

7.7.4 *Benefit-to-Cost Ratio*

The benefit-to-cost ratio is a traditional indicator of cost-effectiveness in which benefits and costs are measured in dollars and their ratio obtained. The analyst calculated the ratio for each strategy, and higher ratios imply higher cost-effectiveness. In this study, the benefit was safety improvement (i.e., mean reduction in frequency of LOV) of a strategy weighted by the replacement cost of an orbiter. That is,

$$\text{benefit} = (\text{mean baseline frequency of LOV} - \text{mean strategy frequency of LOV}) \times (\text{orbiter replacement cost}) \quad (7\text{-}1)$$

Benefit was expressed in this case in units of dollars per mission. The analyst used an orbiter replacement cost of $2.5 billion (U.S. Congress 1989). Cost, in the benefit-to-cost ratio, was the out-of-pocket expenditures of each strategy expressed in dollars per flight by amortizing over the expected number of flights remaining in the shuttle program. The analyst applied an amortization factor (AF), such that

$$\text{cost per flight} = (\text{mean cost from Figure 7-5}) \times (\text{AF}) \tag{7-2}$$

where,

$$AF = \frac{i}{1 - \left(\frac{1}{i+1}\right)^r} \tag{7-3}$$

In Equation 7-3, the analyst assumed that an individual orbiter would fly about three missions per year for the next 20 years (U.S. Congress 1989). The interest, i, was assumed to be 9% per year. The analyst used an amortization period, r (composed of 60 four-month intervals) to obtain AF = 0.036. The decision rule was to select the strategy that maximizes the ratio of the benefit of Equation 7-1 to the cost of Equation 7-2.

Figure 7-6 shows the benefit-to-cost ratios of each strategy. Under the constraints of safety and cost as the only criteria, we can clearly see the ranking of the strategies. The high cost of Strategy F makes it appear less cost-effective than any of the other strategies. Strategy C appears by this indicator to be the most cost-effective because its safety improvement is comparable to that of the other strategies and its cost is approximately half that of the nearest competitor. For comparable safety benefits, applying the benefit-to-cost ratio results in the least-cost option. The expected impact indicator that I discuss in the next section provides a somewhat different perspective.

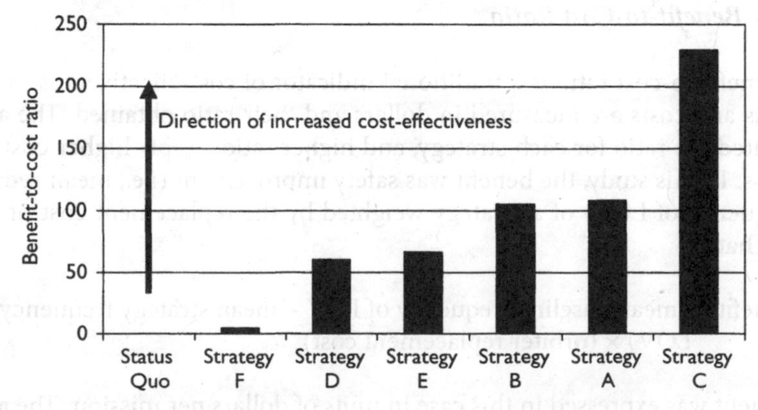

Figure 7-6. ***Benefit-to-Cost Ratios of the Proposed Strategies***

7.7.5 *Expected Impact*

The expected impact, an indicator arising from value-impact assessment, is a generalization of cost-benefit analysis (Nelson and Kastenberg 1986). Typically, the expected impact, I, is the total loss associated with all criteria. In this example, it was the sum of the amortized out-of-pocket expenditures and the total loss (L_T) associated with shuttle LOV resulting from APU-triggered events. L_T is as follows:

$$L_T = (\text{frequency of LOV of a strategy}) \times (\text{orbiter replacement cost}) \quad (7\text{-}4)$$

and the expected impact, I, is as follows:

$$I = \text{cost per flight (from Equation 7-2)} + L_T \quad (7\text{-}5)$$

The decision rule was to minimize I. Figure 7-7 shows the expected impact of each strategy and that of the status quo (the baseline orbiter). We can see that all strategies are an improvement over the status quo and all have similar net impacts. The similarity stems from the term L_T, which has a small variability over the strategies, and is much larger than the cost per flight. Replacing an orbiter is extraordinarily expensive. Therefore, even the relatively small LOV frequencies among the strategies, when weighted by the large cost of orbiter replacement, yield large L_T terms relative to the cost-per-flight terms. The expected impacts of the strategies do not differ significantly because none of the strategies have significantly different LOV frequencies. By this indicator, Strategies B, D, and E appear to be the most desirable because they have a slightly lower LOV frequency than Strategy C. The very high vehicle replacement cost gives a very high weight to small changes in LOV frequency. The out-of-pocket cost of Strategy F is large

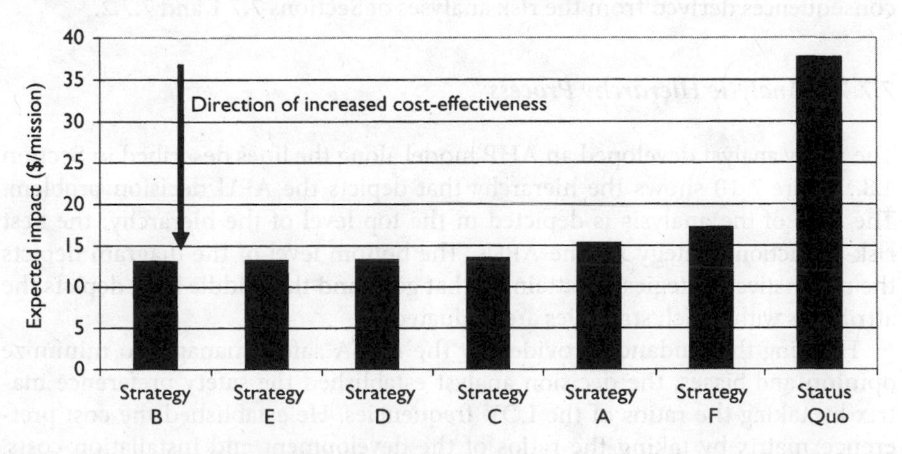

Figure 7-7. *Expected Impact of Each Strategy*

enough, however, to compensate for its somewhat lower LOV frequency. Finally, Strategy F still appears to be the least desirable (except for the status quo).

For this example, the expected impact indicator was heavily weighted toward safety improvement, whereas the benefit-to-cost ratio was weighted toward lower out-of-pocket cost. Two traditionally accepted indicators yielded somewhat different perspectives and apparent recommendations about the strategies.

7.7.6 *Decision Tree Analysis*

Figure 7-8 shows a decision tree using dollars as a common metric. Each path from left to right of the tree depicts a prospective choice (following the square decision node) and chance-driven outcome (following the round chance node). As before, safety is converted into replacement cost. The decision nodes show the total development and installation costs and the chance nodes show the probability of LOV and the replacement cost. The analyst evaluated the cost of each path in the decision tree, including the expected cost of the lotteries defined by the chance nodes (see Section 4.5.2) and the out-of-pocket costs. The decision rule was to minimize the expected cost. In this case, the recommended selection is shown in the figure as indicated by a TRUE for Strategy C followed by Strategies B, A, E, D, status quo, and F, in that order. Again, we see that changing the decision rule changes the rankings, although this ranking is similar to that of the benefit-to-cost ratio.

The analyses described up to now used dollars as a common metric of comparison (by converting safety into an estimated form of dollars). Another, perhaps more illuminating perspective comes from multiattribute decision theory (described in Chapter 4), in which analysts can treat safety in a nonmonetary context and retain cost within its natural monetary context. Here, the analyst seeks to learn, "What is the best option when we consider both attributes?" Analogous to Figure 4-7, the decision tree model for the multiattribute decision analysis in this study is shown in Figure 7-9. This depicts both the cost and safety consequences derived from the risk analyses of Sections 7.7.1 and 7.7.2.

7.7.7 *Analytic Hierarchy Process*

The study analyst developed an AHP model along the lines described in Section 4.8. Figure 7-10 shows the hierarchy that depicts the APU decision problem. The goal of the analysis is depicted in the top level of the hierarchy: the best risk-reduction strategy for the APUs. The bottom level of the diagram depicts the alternative strategies for attaining that goal, and the middle level depicts the attributes with which strategies are evaluated.

Heeding the guidance provided by the NASA safety manager to minimize opinion and biases, the decision analyst established the safety preference matrix by taking the ratios of the LOV frequencies. He established the cost preference matrix by taking the ratios of the development and installation costs. The LOV frequencies and costs of Figure 7-9 were used to develop the pairwise

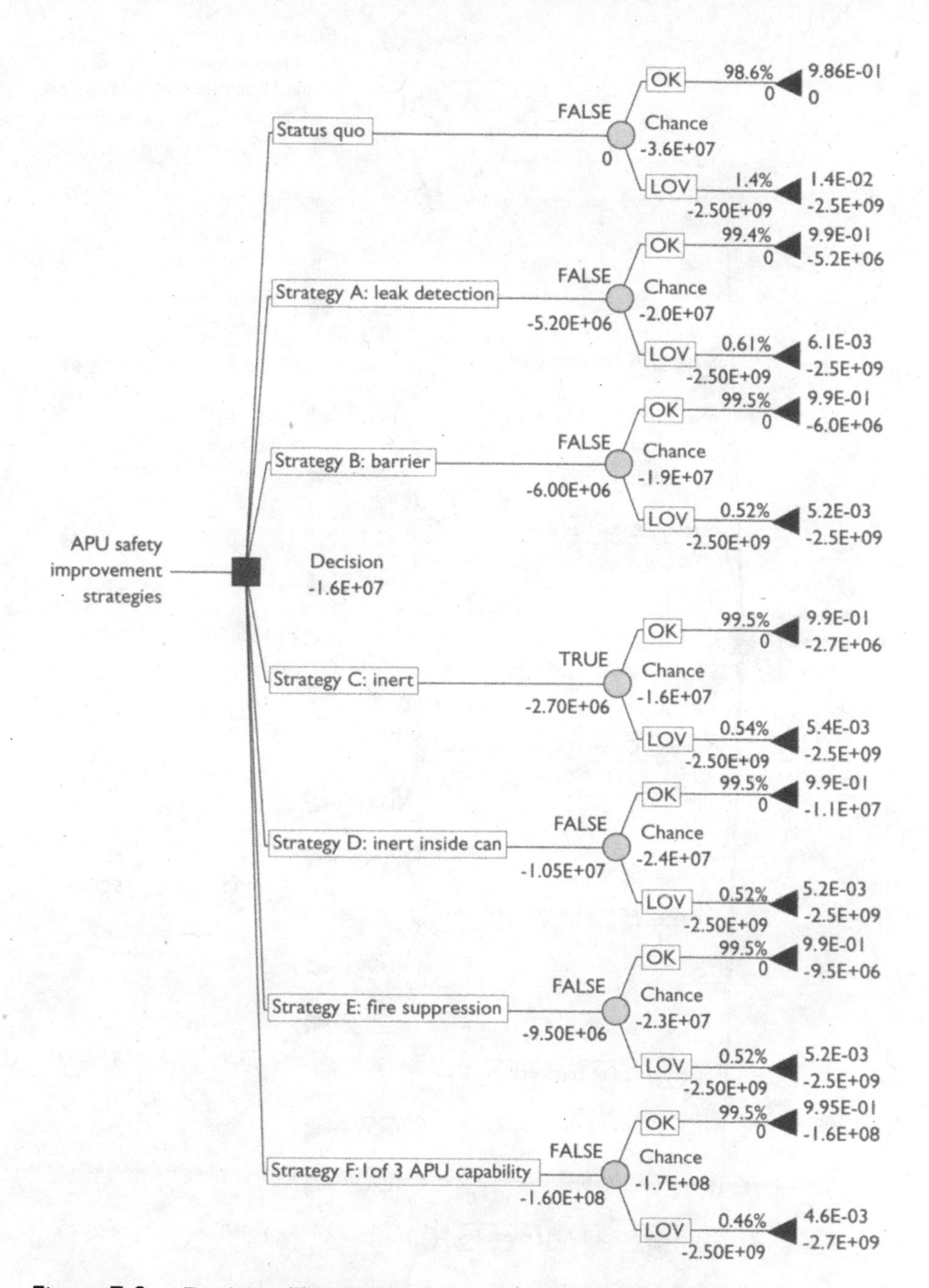

Figure 7-8. ***Decision Tree Using Cost as the Common Metric***

Notes: APU = auxiliary power unit; LOV = loss of vehicle.

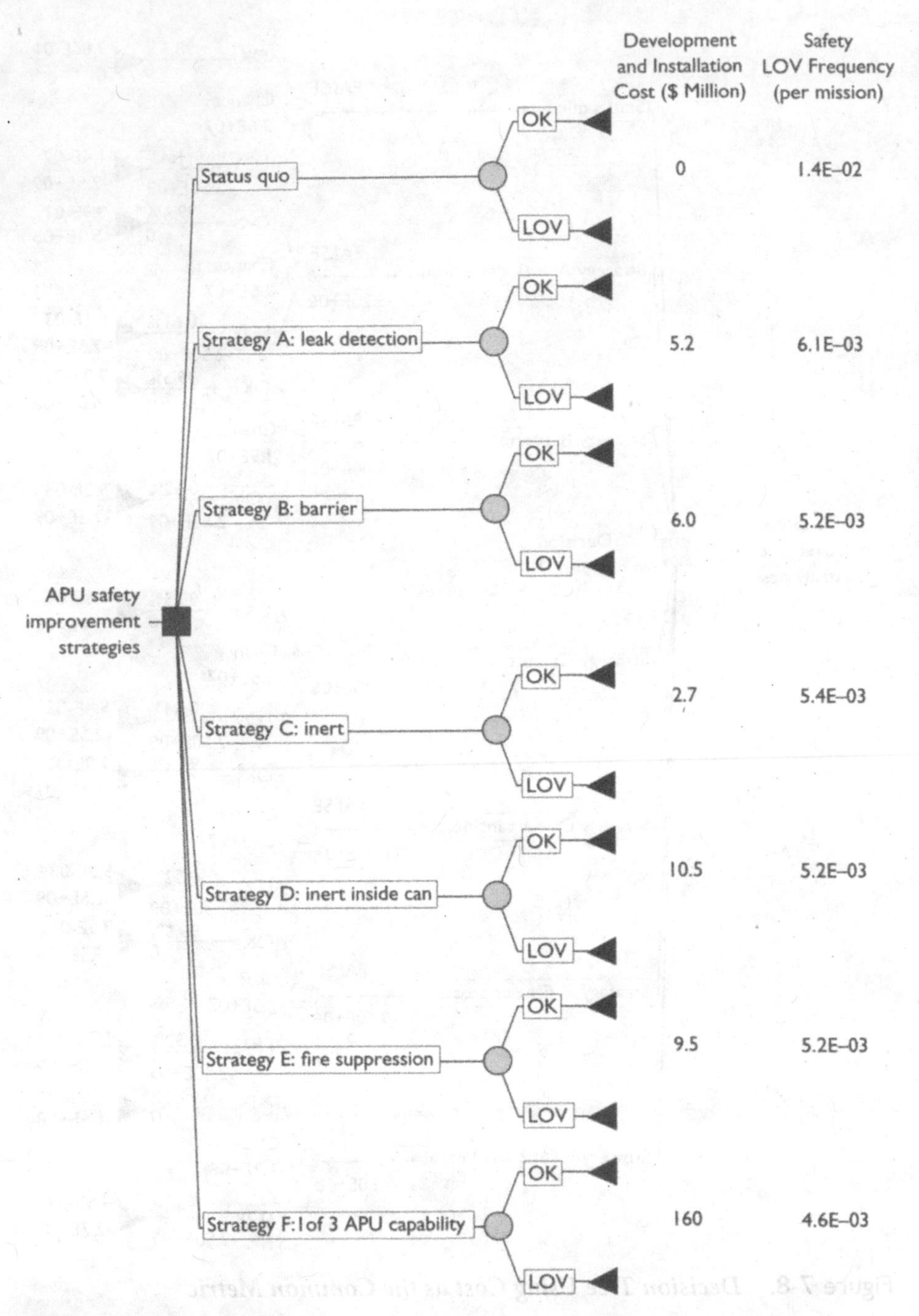

Figure 7-9. ***Multiattribute (Cost and Safety) Decision Tree Model***

Notes: APU = auxiliary power unit; LOV = loss of vehicle.

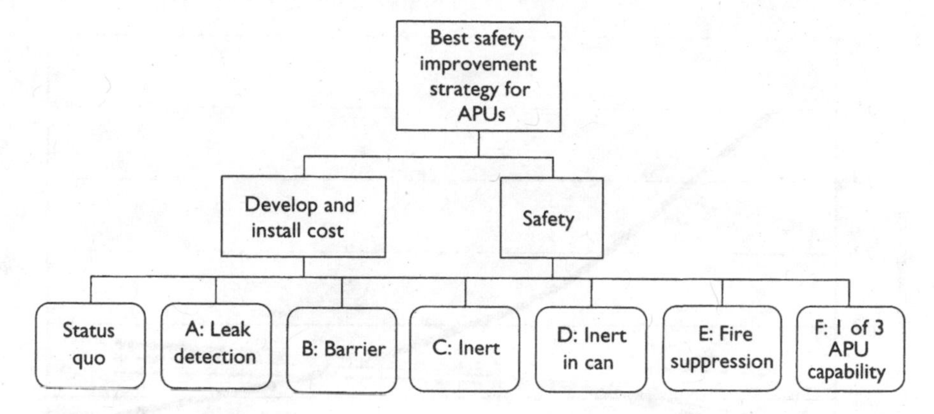

Figure 7-10. ***Hierarchy for Auxiliary Power Unit (APU) Safety Improvement Strategies***

comparisons for this evaluation. Solving the matrices led to the principal eigenvectors (i.e., the normalized ranking with respect to cost and the normalized ranking with respect to safety) shown in Table 7-1. The assumption of no development and installation costs for the status quo led to its very high cost ranking. Its higher LOV frequency led to its low safety ranking.

The matrix constructed from these eigenvectors is weighted by a safety versus cost preference matrix to obtain the rankings of the strategies without utility theory. The analyst developed preliminary decision trajectories, shown in Figure 7-11, to present the results without utility functions.

In this figure, we can see a safety importance weight, $w_s = 0.5$, indicating that cost and safety are equally important. Note that as cost decreases in importance with respect to safety, from the DM's viewpoint, the status quo dramatically decreases in ranking. As a safety importance weight increases beyond 0.6., all strategies become preferred to the status quo and there is little distinction among

Table 7-1. ***Analytic Hierarchy Process (AHP) Principal Eigenvectors of Preference Matrices***

Alternative	*Normalized cost ranking*	*Normalized safety ranking*
Status quo	0.65	0.06
A: Leak detection	0.07	0.14
B: Barrier	0.06	0.16
C: Inert	0.13	0.15
D: Inert inside can	0.04	0.16
E: Fire suppression	0.05	0.16
F: One of three APU capability	0.003	0.18

Note: For auxiliary power unit (APU) safety improvement—no utility functions.

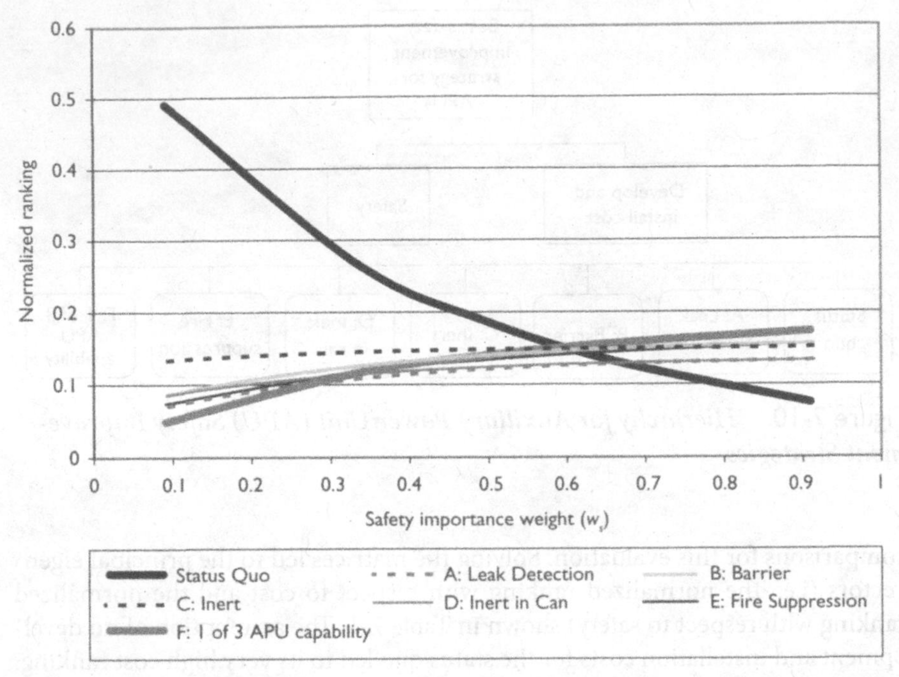

Figure 7-11. ***Decision Trajectories from the Analytic Hierarchy Process (AHP)***
Note: Without utility functions for auxiliary power unit (APU) safety improvement strategies.

Strategies B, C, D, and E. In the extreme, when cost is considered unimportant, the most expensive strategy (F) is preferred because it has the lowest LOV frequency. Strategy A and the status quo, though, are clearly less preferred. Other than the status quo, Strategy C, the intuitive choice, is preferred for the region in which $w_s < 0.6$. This brings to light one of the assumptions the analyst made during the intuitive selection—that strategies A, B, C, D, and E have essentially the same LOV frequency. The decision trajectory shows the regions over which that assumption is valid and invalid.

7.8 The Value Model and the Alternative Rankings

In this section I describe the use of utility functions for creating the value models using the AHP and decision tree models. This decision analysis required a utility function for cost and one for safety. Although the DM wanted to minimize the use of utility functions, the decision analyst explained that some guidance is needed with respect to reasonable bounds of cost and safety improvement. The DM then said that he would consider any APU alternative that cost more than

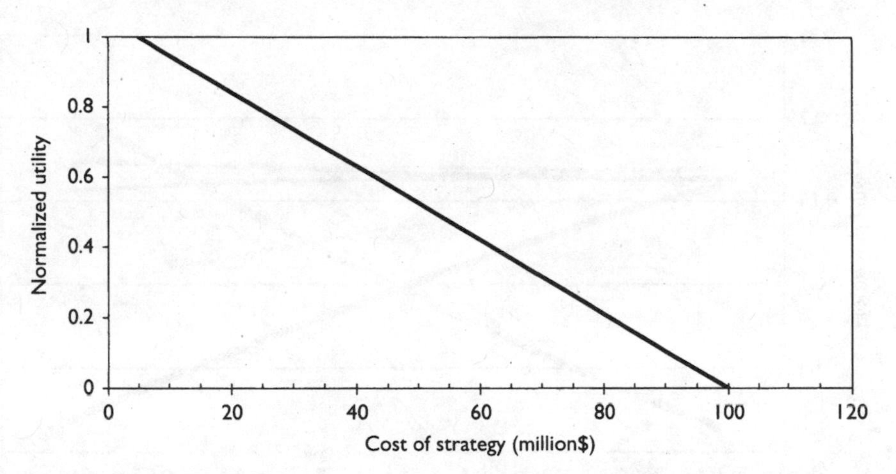

Figure 7-12. *Cost Utility Function for Auxiliary Power Unit (APU) Safety Improvement Decision*

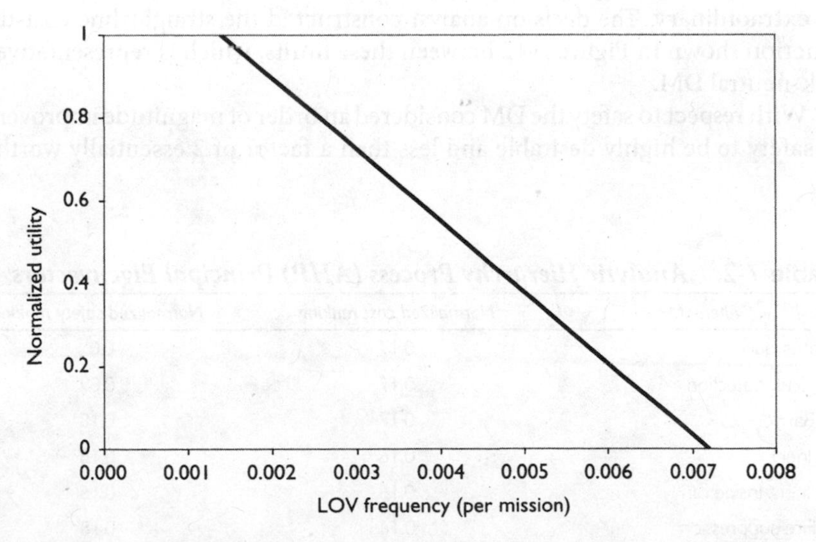

Figure 7-13. *Safety Utility Function for Auxiliary Power Unit (APU) Safety Improvement Decision*

Note: LOV = loss of vehicle.

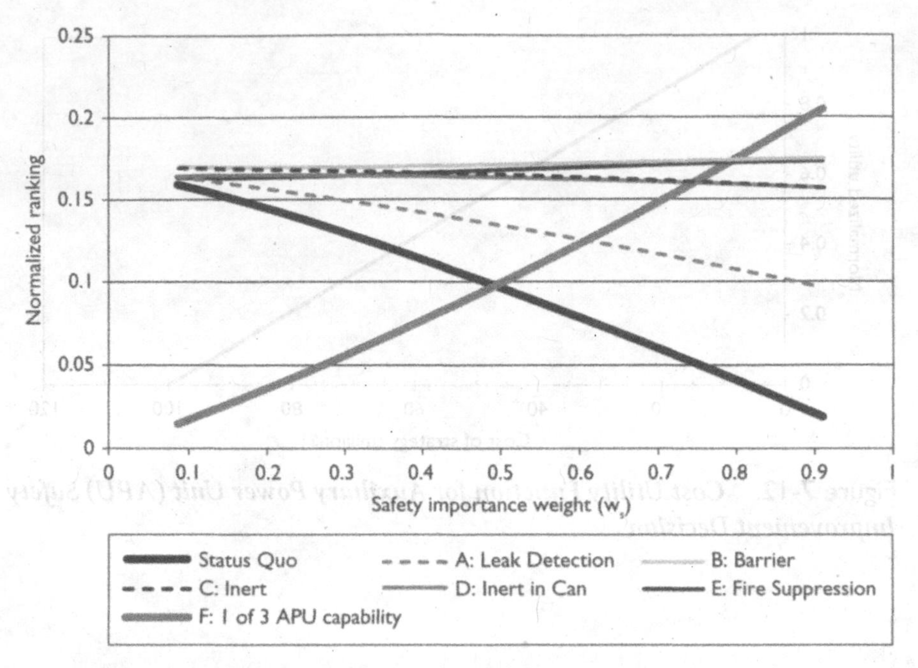

Figure 7-14. *Decision Trajectories from Analytic Hierarchy Process (AHP)*

Note: With utility functions for auxiliary power unit (APU) safety improvement strategies.

$100 million to be outrageously expensive and any that cost under $5 million to be extraordinary. The decision analyst constructed the straight-line cost-utility function shown in Figure 7-12 between these limits, which is representative of a risk-neutral DM.

With respect to safety the DM considered an order of magnitude improvement in safety to be highly desirable and less than a factor of 2 essentially worthless.

Table 7-2. *Analytic Hierarchy Process (AHP) Principal Eigenvectors*

Alternatives	*Normalized cost ranking*	*Normalized safety ranking*
Status quo	0.17	0.0
A: Leak detection	0.17	0.09
B: Barrier	0.17	0.18
C: Inert	0.16	0.18
D: Inert inside can	0.16	0.16
E: Fire suppression	0.16	0.18
F: One of three APU capability	0.0	0.18

Notes: Of preference matrix for APU safety improvement—with utility functions. Using DM's preferences via the utility functions.

Noting that the current LOV frequency was 0.014, the analyst constructed the risk-neutral safety utility function shown in Figure 7-13.

In the AHP, the values derived from the utility function substitute for the calculated consequences when determining the ranking of alternatives (see Section 4.6). In this example, the analyst began by using the calculated consequences in Figure 7-9 and the DM's utility functions for safety and cost, which are shown in Figures 7-12 and 7-13. Using the relationship between consequences and utilities in these figures, the consequences were converted to their equivalent utilities. These utility values were all the analyst needed to create paired comparisons and then to construct a preference matrix composed of the eigenvectors for each attribute. This process was just like the AHP described in Section 7.7, but instead of using the ratio of consequences from Figure 7-9, the analyst used the ratio of the equivalent utilities. Table 7-2 is the resulting preference matrix.

Strategy F had no value to the DM because it cost more than $100 million. All the other costs were, in the DM's perspective, of similar value. With respect to safety, the status quo had no value because the DM's objective was to improve safety. Strategy A was only slightly below the DM's cutoff. From the DM's perspective, the rest were of nearly equal value.

The remainder of the AHP is just like that discussed in Section 4.8.3. A matrix of eigenvectors was developed from the relative rankings with respect to each attribute; an eigenvector of the cost versus safety relative importance ratios was developed; and the dot product of the two was taken to obtain the desired alternative rankings for each value of w_s to create decision trajectories.

Figure 7-14 shows how the preferences were reflected in the decision trajectories.

In looking at the figure, we can see that as the safety importance weight, w_s, increases, preference goes to the options with the lowest LOV frequency, irrespective of cost. It appears that Strategies B, D, and E are essentially equally good choices over the range of w_s. The lowest cost strategy, C, is preferred at low values of w_s. Recall that the intuitive assessment suggested that Strategy C was the best because it cost the least and had nearly the same LOV frequency. The decision trajectories make that judgment explicit. The intuitive judgment is valid for low assignment of safety importance. If safety importance has a high value, even small differences affect the rankings.

With the objective consequences in the decision tree of Figure 7-9 and the utility functions of Figures 7-12 and 7-13, the analyst could substitute utilities for consequences. Figure 7-15 shows the decision tree with that substitution. Except for Strategy F and the status quo, the figure shows little variability across the cost or safety utilities. This is similar to the conclusion drawn from the preference matrix of Table 7-2. The final analysis performed by the decision analyst was to apply Equation 4-11 with varying values of w_s for each strategy. The resultant decision trajectory is shown in Figure 7-16.

Normative decision analysis uses simple linear weighting to add the safety and cost utility values (via Equation 4-11). The AHP uses an entirely different matrix solution. This accounts for the different slope of the decision trajectories when comparing Figure 7-16 with Figure 7-14. Nevertheless, the overall rankings have

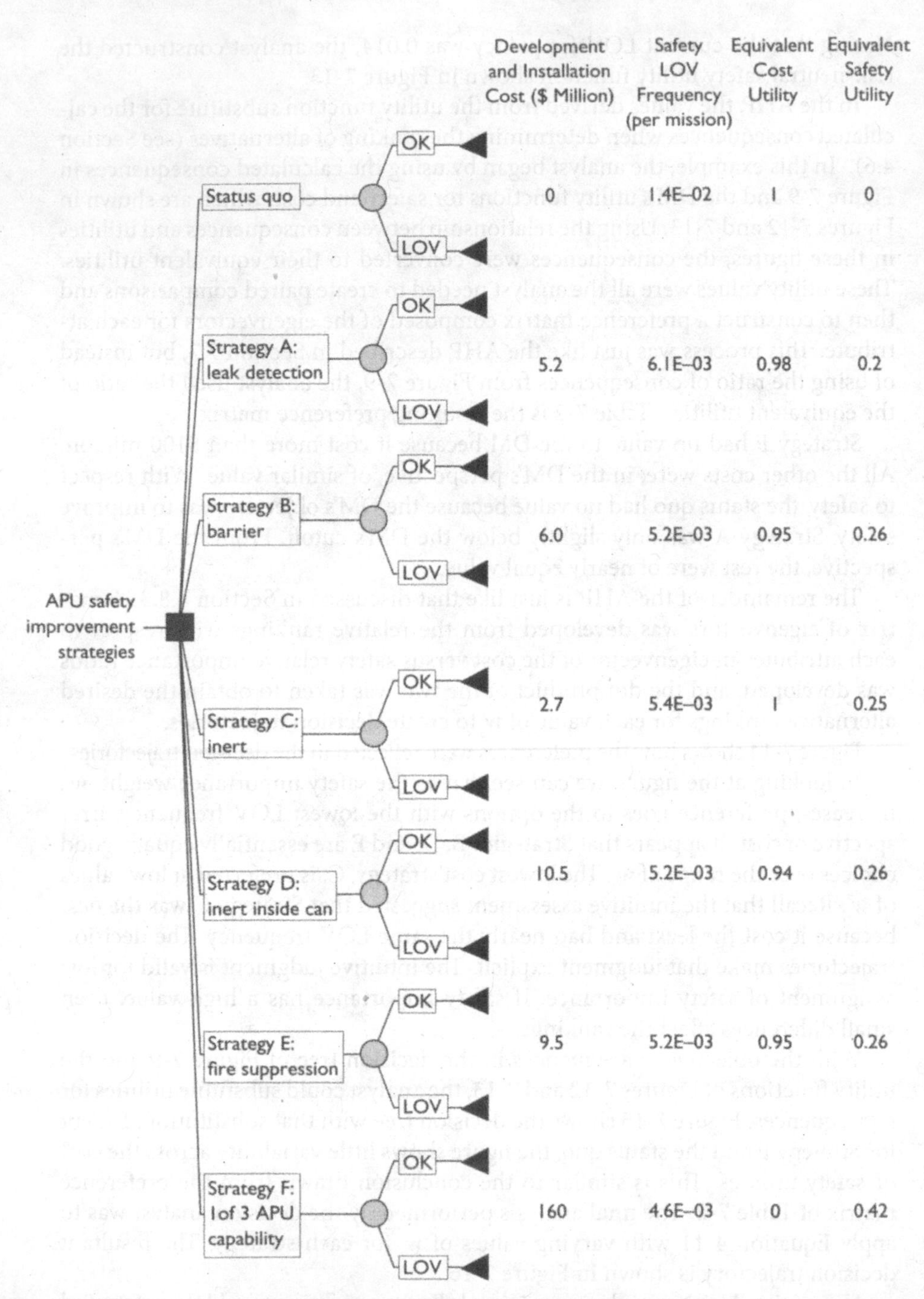

Figure 7-15. ***Decision Tree for Auxiliary Power Unit (APU) Safety Improvement***

Notes: Showing consequences and utilities. LOV = loss of vehicle.

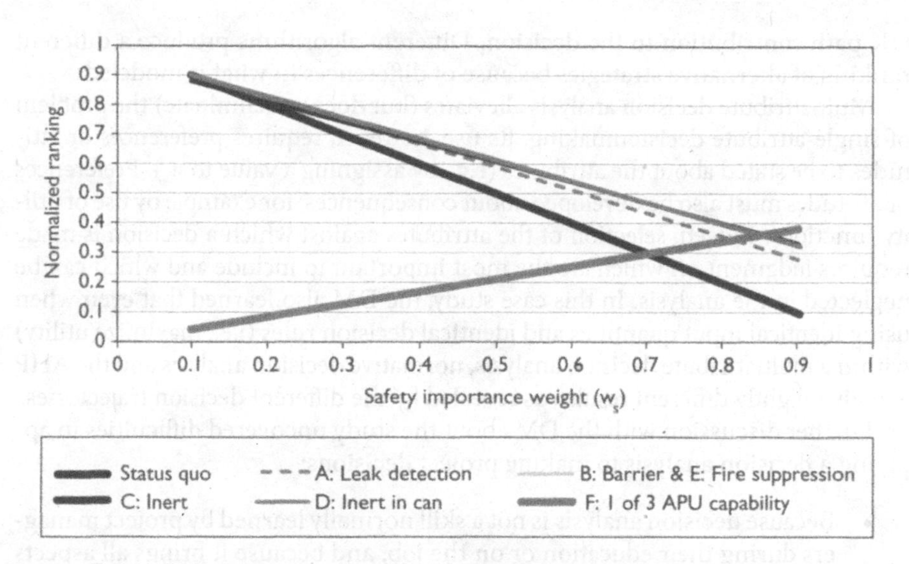

Figure 7-16. ***Decision Trajectories from Decision Tree***

Note: With utility functions for auxiliary power unit (APU) safety improvement strategies.

the same trend and are similar. Strategy C is preferred for low values of w_s and Strategies B, E, D, and F are preferred for higher values of w_s.

7.9 Observations about the Risk Assessment Process in the APU Study

What was the DM's reaction to this demonstration study? He had wanted an objective analysis that would mitigate the effects of biases, judgments, attitudes, and opinion. Although the formal, structured decision analysis had the desired mitigating effect, the DM in this study learned that no decision can be free from these. In the simplest cases of decision analysis in which money is the common metric, for example, the DM is (perhaps unwittingly) making a judgment that one attribute—cost expressed in dollars—is sufficient for the decision, even with all the other factors such as safety, reliability, injury, fatalities, time, and prestige in play.

Attitudes are also inherent in the choice of decision model or decision rule. Using the benefit-to-cost ratio implies that the DM prefers to weigh costs against benefits.

In contrast, using expected impact implies that the DM simply wants to minimize costs. The decision tree of Figure 7-8 is based solely on cost, but in contrast to expected impact, Figure 7-8 takes into account probabilities that lead to success (i.e., ending OK) as well as those ending in LOV. The decision tree minimizes the expectation value considering the lottery of both success and failure. The expected impact algorithm used here, on the other hand, does not include the

OK path contribution to the decision. Different algorithms produce a different ranking of alternative strategies because of differences in what is modeled.

Multiattribute decision analysis alleviates (but does not eliminate) the problem of single-attribute decisionmaking. Its use, however, requires preferences or attitudes to be stated about the attributes (e.g., by assigning a value to w_s). Preferences or attitudes must also be developed about consequences, for example by use of utility functions. Indeed, selection of the attributes against which a decision is made requires judgment on which are the most important to include and which can be neglected in the analysis. In this case study, the DM also learned that even when using identical input quantities and identical decision rules (i.e., maximize utility) within a multiattribute decision analysis, normative decision analysis and the AHP can give slightly different results, as revealed by the different decision trajectories.

Further discussion with the DM about the study uncovered difficulties in applying a decision analysis to making project decisions:

- Because decision analysis is not a skill normally learned by project managers during their education or on the job, and because it brings all aspects of the decision into the light, DMs need time to become comfortable with recommendations emerging from the process.
- Conducting a decision analysis would require hiring someone with the requisite skills. This may involve reallocating project expenditures.

The intuitive approach has two problems. First, it uses inherent judgments about decision rules and assumptions about preferences. Second, neither of these are explicit. This means that the underlying DM's attitudes—which led to a choice of one alternative over another—cannot be examined for consistency with the DM's actual preferences. In this case study, it appeared that Strategy C was the best choice from an intuitive perspective. This arose from the judgments that its safety was not significantly different from the others and its cost was lowest. Both the AHP and decision tree multiattribute analyses, however, explicitly used the actual calculated mean safety and cost consequences of each strategy. The intuitive approach turned out to be valid only if safety was given a relatively low weight. Although the DM had not done this on purpose, it was the result of the implicit reasoning. Using a formal decision analysis contributes to making explicit the attitudes, opinions, assumptions, and preferences that go into a decision.

Note

1. Section 3.3 describes and provides an example of the development of a risk model using ETs and FTs.

References

MDAC (McDonnell Douglas Astronautics Company). 1987. *Space Shuttle Probabilistic Risk Assessment Proof-of-Concept Study Volume III.* WP NO. 1.0-WP-VA88004-03. Houston, TX: MDAC.

NASA. 1993. *Space Shuttle Operational Flight Rules, National Aeronautics and Space Administration, Mission Operations Directorate.* NSTS-12820, PCN-19. Houston, TX: Johnson Space Center.

Nelson, P.F., and W.E. Kastenberg. 1986. An Extended Value–Impact Approach for Nuclear Regulatory Decision-making. In *Nuclear Engineering and Design*, Vol. 93. New York: Elsevier, 311–317.

U.S. Congress. 1989. *Round Trip to Orbit: Human Spaceflight Alternatives—Special Report.* OTA-ISC-419. Washington, DC: U.S. Congress Office of Technology Assessment.

CHAPTER 8

The Decision to Launch the Cassini Spacecraft

TYPICALLY, LONG-DURATION SPACE MISSIONS to the outer planets require the use of nuclear-powered electrical and heat sources for the spacecraft. The *Cassini* spacecraft (launched on the *Titan IV/Centaur* launch vehicle) carries three radioisotope thermal-electric generators (RTGs), as illustrated in Figure 8-1. An RTG, pictured in cutaway in Figure 8-2, converts the energy from nuclear decay (predominately alpha decay) into heat and then electricity. As shown in Figure 8-2, an RTG has an aluminum housing that contains (1) heat-source modules of radioactive plutonium dioxide (PuO_2) surrounded by graphite and (2) components needed for thermoelectric conversion of energy generated by decay of the radioactive PuO_2. An RTG is generally cylindrical, with a diameter of about 0.43 m and a length of about 1.17 m.

The radioactive material is contained within 18 graphite "bricks" called general-purpose heat-source (GPHS) modules.[1] Each module contains about 604 g of fuel and has roughly the dimensions of a stack of five plastic CD cases. The outer portion of the module is a graphite-weave aeroshell. Within the aeroshell are two graphite-impact shells (GISs), one of which is depicted in Figure 8-3. Two small iridium-clad fuel cylinders—one of which is shown in the figure—reside within each GIS. Each fuel cylinder contains about 151 g of fuel.[2] About 71% (by weight) of that fuel is Pu-238. The remainder is composed of other plutonium isotopes (including 239 through 242), actinides other than plutonium actinides, and oxygen (which is chemically bound to the heavy metal to form the PuO_2).

RTGs are used for extended solar system missions where the energy requirements exceed the capability of solar power. Typically, an RTG contains about 130,000 curies of plutonium.

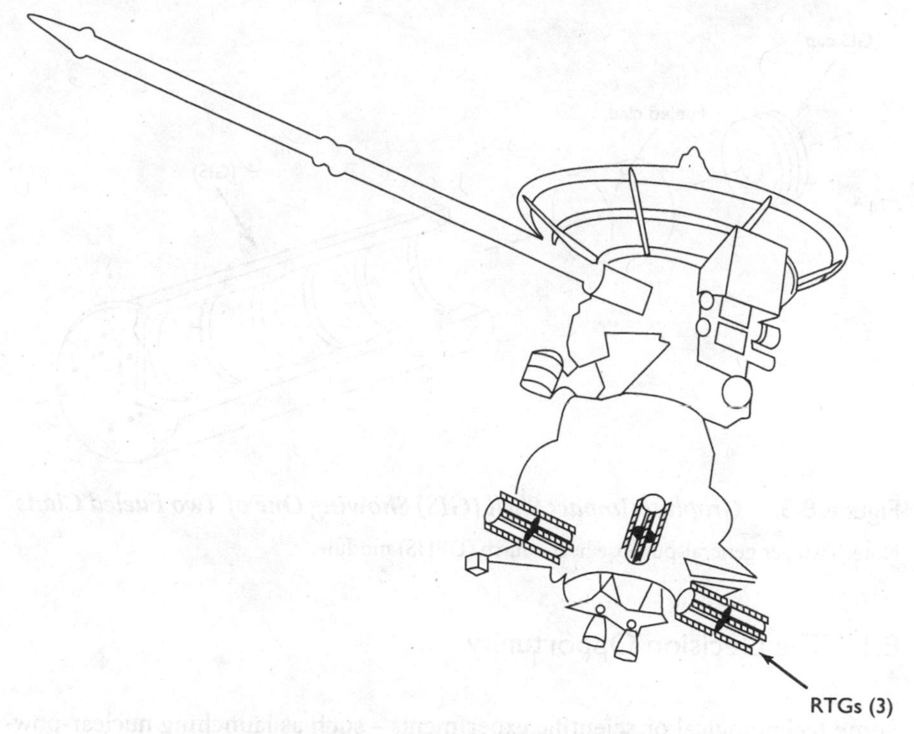

Figure 8-1. ***Outline of the* Cassini *Spacecraft Showing Three Radioisotope Thermal-Electric Generators (RTGs)***

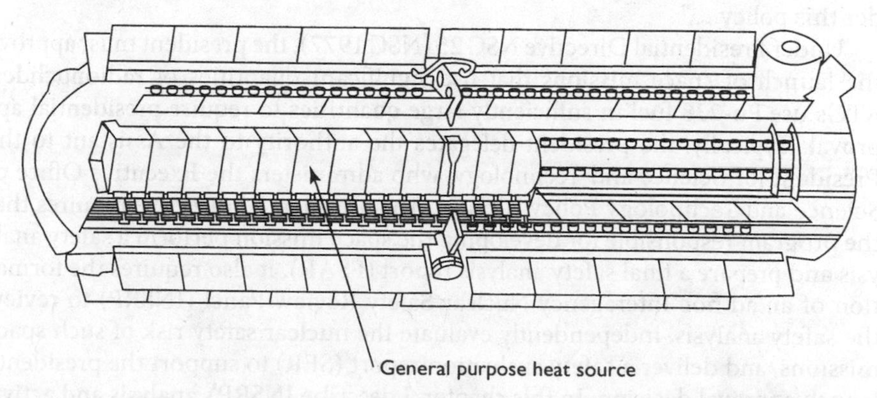

Figure 8-2. ***Radioisotope Thermal-Electric Generator (RTG) Cutaway***

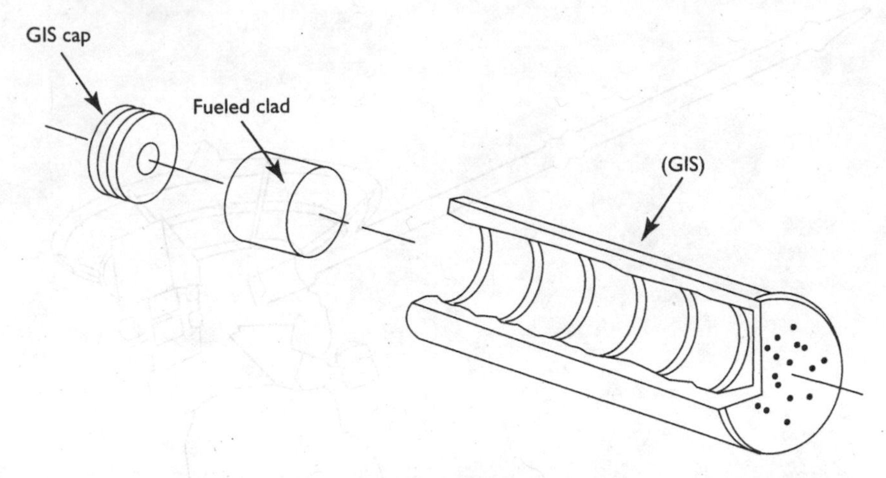

Figure 8-3. ***Graphite-Impact Shell (GIS) Showing One of Two Fueled Clads***

Note: Two per general-purpose heat-source (GPHS) module.

8.1 The Decision Opportunity

Some technological or scientific experiments—such as launching nuclear-powered systems into space—might have a large-scale or protracted effect on the world's physical or biological environment. The Executive Office of the President of the United States formally recognized this issue early in the 1960s (NSAM 1963, 1965). A comprehensive directive that establishes the president as the decisionmaker (DM) for launching of nuclear devices was promulgated in 1977 and revised in 1996 (NSC 1977). The directive states, "…experiments which by their nature could reasonably be expected to result in domestic or foreign allegations that they might have major and protracted effects on the physical or biological environment, or other areas of public or private interest, are to be included under this policy…."

Under Presidential Directive NSC 25 (NSC 1977), the president must approve the launch of space missions that use significant quantities of radionuclides. RTGs use Pu-238 fuel in sufficiently large quantities to require presidential approval. Typically, the president delegates the authority to the Assistant to the President for Science and Technology, who administers the Executive Office of Science and Technology Policy (OSTP). The presidential directive requires that the program responsible for developing the space mission perform a safety analysis and prepare a final safety analysis report (FSAR). It also requires the formation of an ad hoc Interagency Nuclear Safety Review Panel (INSRP) to review the safety analysis, independently evaluate the nuclear safety risk of such space missions, and deliver a safety evaluation report (SER) to support the president's launch approval decision. In this chapter, I describe INSRP's analysis and activities for the *Cassini* mission.

8.2 The Problem Statement

In effect, the presidential directive created the following problem statement, phrased as a question: What is the nuclear risk to the environment and biosphere of launching the *Cassini* spacecraft? In this statement, risk is calculated as probability of latent cancer fatalities (LCFs) due to accidental release of plutonium from the spacecraft.

The *Cassini* program performed an extensive quantitative risk assessment, which was coordinated by the *Cassini* Program Office at NASA and the U.S. Department of Energy (DOE) and documented in a FSAR (Lockheed Martin 1997). The FSAR described the safety risks associated with launch, orbit, and earth-gravity-assist (EGA)[3] maneuvers of the *Titan IV/Centaur/Cassini* vehicle. The FSAR presented a detailed description of the models, methods, assumptions, data, calculations, uncertainties, and risk results. Major technical contributors to the FSAR were NASA's Jet Propulsion Laboratory (JPL), Johns Hopkins Applied Physics Laboratory, Sandia National Laboratory (SNL), and Lockheed Martin Corporation.

After reviewing the *Cassini* FSAR, the *Cassini* INSRP provided a SER to the OSTP. INSRP provided review and feedback to the *Cassini* program during the development of the FSAR and then performed an independent quantitative risk assessment.

By performing reviews for about 30 years, the INSRP had learned that the RTG design was robust against both launch and reentry accidents. The nuclear-related risks were found to be exceedingly low. For example, for the *Cassini* mission, the frequency of a single latent cancer fatality or more was estimated to be between one in 2,500 and one in 10,000,000 missions, with a median of approximately one in 30,000 missions.

8.3 The Objectives and the Attributes

It was clear from the presidential directive that safety was the single decision attribute. The scope of work included performing a probabilistic risk assessment (PRA) to obtain the probability distribution of LCFs. The results were represented as complementary cumulative distribution functions (CCDFs) with uncertainties (see Sections 2.2 and 2.3). Concerns other than safety were implied but not stated in the directive, including (1) national security concerns stemming from a launch accident that would cause U.S. technology to land on foreign soil; (2) political concerns arising from allowing a nuclear system to fly over a foreign country; and (3) environmental concerns stemming from the contamination of agricultural land after a severe accident.

At the root of these, however, was safety. Minimizing accidental releases and mitigating the effects of such accidents would ameliorate the other concerns. Cost was not a decision attribute, but a constraint. Each such project is funded with a budget, and cost is not important as long as the project stays within that budget. Exceeding the budget would lead either to project cancellation or an increased budget. Because of the prime importance of safety, budget problems are not allowed to lead to safety problems.

8.4 The Alternatives

During the development of any spacecraft with nuclear-powered systems and its launch vehicle, the project team develops an environmental impact statement (EIS). This statement sets out the justification for the mission and discusses the pros and cons of alternatives for achieving the mission objectives. The team considers costs and safety risks, among other factors. The EIS for the *Cassini* mission successfully justified going ahead with spacecraft and launch vehicle development for the purpose of increasing scientific knowledge about Saturn (NASA 1995, 1997). By the time NSC-25 became applicable, the launch vehicle and spacecraft were being developed. By the time of the presidential decision, both vehicles had been built. As a result, only two alternatives remained: launch or do not launch.

8.5 The Risk Analysis Overview

The assessment of risk, including uncertainties, associated with postulated scenarios potentially capable of releasing radioactive material during space missions is a complex process involving numerous models, analyses, and experiments. Probabilistic techniques, as well as more traditional engineering analyses and interpretation of data, are typically combined to produce the risk assessment. The complexity of the physical processes and phenomena that are at the focal point of the risk assessment demand that both natural variability of physical processes (aleatory uncertainties) and the uncertainties in knowledge of these processes (epistemic uncertainties) are modeled.

INSRP's risk assessment was divided into seven elements:

1. Identification of launch accident scenarios and the calculation of frequency of these scenarios
2. Identification of reentry accident scenarios either from the Earth's orbit or from EGA fly-bys and the calculation of frequency of these scenarios
3. Development of the adverse environments resulting from launch accident and reentry accident scenarios that have an impact on the RTGs
4. Assessment of damage to the RTGs and characterization of the released fuel (called the "source term")

5. Assessment of radioactive material dispersion and demography of exposure
6. Assessment of human uptake (defined as the amount of deposited radioactive material that could potentially be ingested by humans after a postulated accident scenario), human health effects, and land contamination and
7. Assessment of epistemic and aleatory uncertainties

I discuss each of these elements in the rest of this section with an emphasis on identification of uncertainties. This risk assessment involved modeling the complex interaction among all events and phenomena that could influence the release, dispersion, and transport of nuclear material. INSRP used a scenario-based approach like the one I described in Chapter 3, in which models and calculations from many different disciplines are used. The overall process of this risk assessment, with its attendant uncertainty analysis, was divided into disciplines that roughly correspond to the division of responsibilities among the INSRP working groups.

8.5.1 Identifying Launch Accident Scenarios

A Launch Abort Working Group was put together to review this area. The group identified accident scenarios that might have the potential for adverse effects on RTGs. The experience of group members with past launch vehicle mishaps greatly aided the development of event trees and the underlying analysis models used to represent the scenarios and their associated physical process. Uncertainties were associated with categorization of scenarios, system and component failure and success data, models, and assumptions.

8.5.2 Identifying Reentry Accident Scenarios

The Reentry Working Group reviewed this area. Reentry accident scenarios fell into two general categories: (1) those that cause reentry of the spacecraft into the Earth's atmosphere from Earth orbit or suborbital altitudes and (2) those that cause reentry owing to mishaps before or during an Earth fly-by. Earth fly-bys occur because planetary gravity assists must be used to achieve sufficient velocities to make interplanetary missions feasible. The group developed a map of possible reentry paths and found that, for such paths, atmospheric releases from orbital or suborbital reentries were not expected to occur. Releases on the ground, however, were possible if the RTG hit a hard surface on Earth. Because high-speed fly-by scenarios can cause atmospheric releases, the group calculated the frequencies—with uncertainty—of such scenarios. The team members used stagnation-point and computational fluid dynamic techniques coupled to heat-balance calculations to investigate the potential conditions in which releases may occur. To assess the likelihood of fly-by reentry and orbital reentry, the group relied on detailed fault trees and calculation of failure scenarios during the spacecraft's flight through the solar system developed by JPL (1993, 1997). In addition to the inherent variability of angle and velocity of reentry, uncertainties were associated with models, experimental data, system and component failure and success data, input variables, and assumptions.

8.5.3 *Developing the Environments*

The Launch Abort Working Group reviewed this area. When launch vehicles fail, various adverse environments surround and impinge on the RTGs. The group used phenomenological analyses backed by experimental data to develop the thermal, overpressure, dynamic pressure, shock, and impact loads on the RTGs from launch accidents. By thoroughly reviewing all available data, the reviewers identified parameters and variables that were important to the determination of fuel release and assessed their uncertainties. In addition to the inherent variability of environments whose parameters were often stochastic, uncertainties were associated with models, input variables, experimental data, and assumptions.

8.5.4 *Characterizing Source Terms*

The Power Systems Working Group reviewed this area. Some accident environments have the potential to breach an RTG's protection and release plutonium fuel. To calculate potential damage to the RTG resulting from accident scenarios, analysts used models of structural, ballistic, impact, shock, thermal, and overpressure behavior. When a release of radioactive material was calculated for a scenario, the analysts obtained the source term. For the *Cassini* mission, the important parameters of the source term were the type of fuel, chemical composition, amount of fuel release, particle size distribution, fraction of fuel vaporized, altitude, velocity, and conditional probability of release. Once a launch or reentry scenario had been defined, INSRP treated the specific insult to the RTG as if it was stochastic. Analysts used Monte Carlo simulation of impingements or impacts on RTGs (resulting from environments) as a means of accounting for the inherent variability of RTG damage and subsequent source terms for each previously identified accident scenario. In addition to the inherent variability of the processes and the source term, uncertainties were associated with models, input variables, experimental data, and assumptions.

8.5.5 *Estimating Radioactive Material Dispersion and Dose*

The Meteorological Working Group reviewed this area, calculating the potential dispersion within the atmosphere and deposition onto the Earth's surface in the event of a release of radioactive materials from RTGs. Postulated scenarios that cause release within the troposphere are subject to wind and other atmospheric conditions at the time of the postulated accident. Because this cannot be predicted in advance, the assessment is inherently probabilistic and, in this case, the group used Monte Carlo techniques to try to capture the variability. Calculating the deposition and resuspension of radioactive material along all potential paths to humans results in the estimated dose to an exposed individual. Overlaying demographic information on the calculated cloud path and deposition yields estimates of overall population doses. In this example, the analysts assumed that scenarios that produce releases above the troposphere would distribute the releases evenly over a more global population. Also considered were reentry scenarios that might lead to ground releases, which might in turn produce local releases to populations along a band of the Earth's surface that corresponds approximately to the orbital path. In addition to the inherent variability of the wind, atmospheric conditions, and demography, uncertainties were associated with models, input variables, experimental data, and assumptions.

8.5.6 Estimating Human Uptake, Health Effects, and Land Contamination

The Biomedical Effects and Environment Working Group reviewed these areas. The group estimated potential levels of human uptake and the amount of ensuing latent cancers in the exposed population resulting from the estimates of the Meteorological Working Group. The stochastic nature of latent health effects implies that for each absorbed dose, there is a conditional probability—with uncertainty—that a fatal cancer might be induced. I must stress that the calculation of latent cancers from small doses does not imply that there will be a demonstrable occurrence of latent cancers. The calculated population exposed and doses considered in the postulated accident scenarios were such that zero cancers cannot be ruled out. For the *Cassini* mission, however, the group also produced estimates of contaminated land areas and cost estimates for decontamination. Such estimates were highly sensitive to policy decisions about de minimus levels and decontamination methods.

In addition to the inherent variability of human uptake, resuspension, and human susceptibility to radiation exposure, the uncertainties in calculated latent cancers were associated with models, input variables, experimental data, and assumptions.

8.5.7 Accounting for Epistemic and Aleatory Uncertainties

The Risk Integration and Uncertainty Working Group reviewed this area. An overall assessment of the risk of the mission with uncertainties identified, quantified, and propagated across all of the calculations was a significant part of INSRP's safety analysis effort. From the review of radiological risk calculations described earlier, the group was able to identify the key aleatory and epistemic uncertainties of the analysis. Analysts used Monte Carlo simulation to obtain an overall quantitative assessment of risk, with aleatory and epistemic uncertainties integrated throughout the analysis. Dominant contributors to these uncertainties were (1) the frequency of fuel releases, (2) the accident environments that affect the RTGs, (3) the amount of fuel released and the fraction of released fuel that is respirable, (4) the paths of dispersion and transport toward population centers, and (5) the resuspension of particles after initial deposition. In addition to the inherent variability of angle and velocity of reentry, uncertainties were associated with models, experimental data, component and system failure and success data, input variables, and assumptions.

8.6 Identifying Three Scenario Categories

Accidents can occur in essentially a continuum of scenarios, each of which might differ slightly from the others. One of the essential features of a risk assessment is the characterization of this continuum into categories of scenarios that can be analyzed without significant loss of information. Potential accident scenarios involving the *Titan IV/Centaur/Cassini* vehicle readily fall into three categories: launch accidents, orbital and suborbital reentry accidents, and accidental reentry of the *Cassini* space vehicle during the EGA maneuver (also called the EGA reentry accident). At the time when the assessment was conducted, the *Titan IV/ Centaur* had a historical failure rate of approximately 10%.

Launch accidents were important to analyze because release of fuel from RTGs might occur in a significant percentage of these accidents, and the fuel could be transported, albeit with low probability, to populations in Florida near the launch site. The dominant launch accident scenario was a malfunction of the *Titan IV/Centaur* that causes an explosion that destroys the vehicle (Figure 8-4).

The explosion generates a massive amount of debris, but about 90% of such accidents would result in the *Cassini* spacecraft, with RTGs attached, remaining nearly intact but separating from the *Titan IV/Centaur* launch vehicle and falling to Earth. The remaining 10% of such scenarios would also destroy the spacecraft, after which time the RTGs and the spacecraft debris would fall to Earth.

I now discuss the most likely of the above two situations. If the intact spacecraft (with the RTGs still attached) were to hit the ground during the first 28 seconds of the mission, the propellant on the *Cassini* spacecraft would explode. In this event, ground impact and explosive forces might cause some of the RTGs to fail and release fuel.

While still in the air or on the ground, the RTGs might also be hit by high-velocity fragments from the exploding vehicle. While on the ground, the RTG housing might be removed by the exploding spacecraft, leaving free modules. The explosion of the spacecraft propellants has the potential to move some of these modules along the ground a short distance. Meanwhile, the solid rocket motor propellant would be broken up during the initial launch vehicle explosion into fragments of various sizes and weights. Should they land on RTG modules, some of the fragments are large enough to damage the modules, releasing fuel.

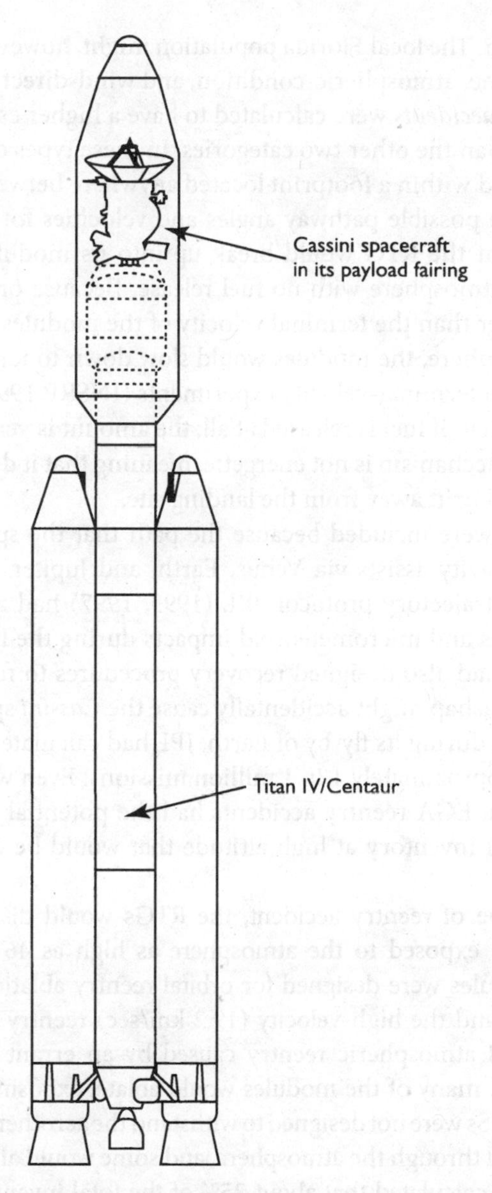

Figure 8-4. *Outline of* Titan IV/Centaur *Launch Vehicle with* Cassini *Spacecraft*

This has a very low probability of occurrence because calculations show that these solid rocket motor fragments tend to hit ground before the spacecraft, and the spacecraft tends to land away from the location of these heavy fragments.

If fuel is released, plume updraft mechanisms would "loft" some of it. The local meteorology is such that many of the lofted fuel particles would enter a mixing layer and be transported either out to sea or over land without interacting with the population. The local Florida population might, however, be affected for some possible plume, atmospheric-condition, and wind-direction scenarios.

Orbital reentry accidents were calculated to have a higher estimated frequency of occurrence than the other two categories. In these types of accidents, RTG modules might land within a footprint located anywhere between 28° north and south latitude. The possible pathway angles and velocities for the *Cassini* mission were such that the RTG would break up into its modules, which would pass through the atmosphere with no fuel release. Because orbital and reentry velocities are higher than the terminal velocity of the modules when they travel through the atmosphere, the modules would slow down to terminal velocity as they fall. Data from terminal-velocity experiments (INSRP 1997) on RTG modules demonstrate that, if fuel is released at all, the amount is very small. Furthermore, the release mechanism is not energetic, meaning that it does not cause fuel particles to loft and drift away from the landing site.

EGA accidents were included because the path that the spacecraft takes to Saturn involves gravity assists via Venus, Earth, and Jupiter. In designing the spacecraft and its trajectory protocol, JPL (1993, 1997) had accounted for internal malfunctions and micrometeoroid impacts during the flight through the solar system. JPL had also designed recovery procedures to further reduce the likelihood that a mishap might accidentally cause the *Cassini* spacecraft to reenter the atmosphere during its fly-by of Earth. JPL had calculated their frequency of occurrence as approximately 1 in 1 million missions. Even with these built-in safeguards, though, EGA reentry accidents had the potential to release a large fraction of the fuel inventory at high altitude that would be dispersed around the world.

During this type of reentry accident, the RTGs would disintegrate and the modules would be exposed to the atmosphere as high as 46 km (150,000 ft). Although the modules were designed for orbital reentry ablation, they were not designed to withstand the high-velocity (19.2 km/sec) reentry that would result from an accidental atmospheric reentry caused by an errant fly-by maneuver. As they descended, many of the modules would ablate, exposing the GISs to the atmosphere. The GISs were not designed to withstand the aerothermal environment of high-speed travel through the atmosphere, and some would allow release of fuel pellets. The INSRP calculated that about 25% of the total inventory of fuel of the three RTGs would be released into the atmosphere at high altitudes. The remainder of the fuel would fall to Earth as chunks or within the GISs, which would offer minimal protection and allow fuel to be released on impact with land.

8.7 Example Risk Assessment for EGA Reentry Scenario

INSRP's risk assessment of this scenario considered the following questions:

- Is the design margin sufficient to allow modules and GISs to survive this high-speed reentry?
- If not, what are the range of fuel releases and their probabilities?

Analyzing the response of the GPHS aeroshells and GISs under this scenario was complicated by having to consider a number of factors:

- The severe reentry conditions
- The predicted high temperatures possible for the reentry
- The aerodynamic loads, which were a factor of 10 higher than for a standard ballistic reentry from Earth orbital velocity
- The uncertainties in the structural and thermal properties
- The modeling uncertainties in thermal response and ablation chemistry of the components

INSRP also found that survival would be a function of the attitude and dynamics of the components during the heating phase. For example, a spinning GIS was less likely to fail than a stable, nonspinning GIS because the material lost to heating would be concentrated in one area in a nonspinning GIS.

Analyses (INSRP 1997; Lockheed Martin 1997) predicted that the GPHS modules would fail structurally for entry flight path angles steeper (more negative) than −16° relative to a vector tangent to the Earth's surface. Such a failure would release the GIS modules into an environment where they would be exposed to severe loads and heating, possibly leading to failure. Failure of a GIS would allow direct aerodynamic heating on the iridium-clad fuel pellets. Reentry heating would cause clads to melt, and much of the fuel would become small particles or vapor. Larger debris pieces, including whole GIS modules, clad fuel pellets, and pieces of the fuel from failed GISs would impact the Earth, some on land and some in water.

INSRP analysts constructed an event tree (Figure 8-5) to delineate reasonable variations of scenarios. Across the top of the figure are the factors that most influence the probability and amount of fuel release. I synopsize these events in the remainder of this section.

Initiating event? The initiating event was accidental reentry of the spacecraft. The dominant causal mechanism of this event would be micrometeoroid impact on a part of the spacecraft, such as a tank, that would produce a change in velocity and direction.

Steep path angle (St)? The distinction between steep and shallow reentry path angles was defined as the demarcation between different aeroshell failure mechanisms. "Steep" reentry path angles were characterized by aerostructural failures that cause the GIS modules to be released during the high heating portion of the trajectory. "Shallow" reentry path angles tend to be characterized by aerothermal ablation failure.

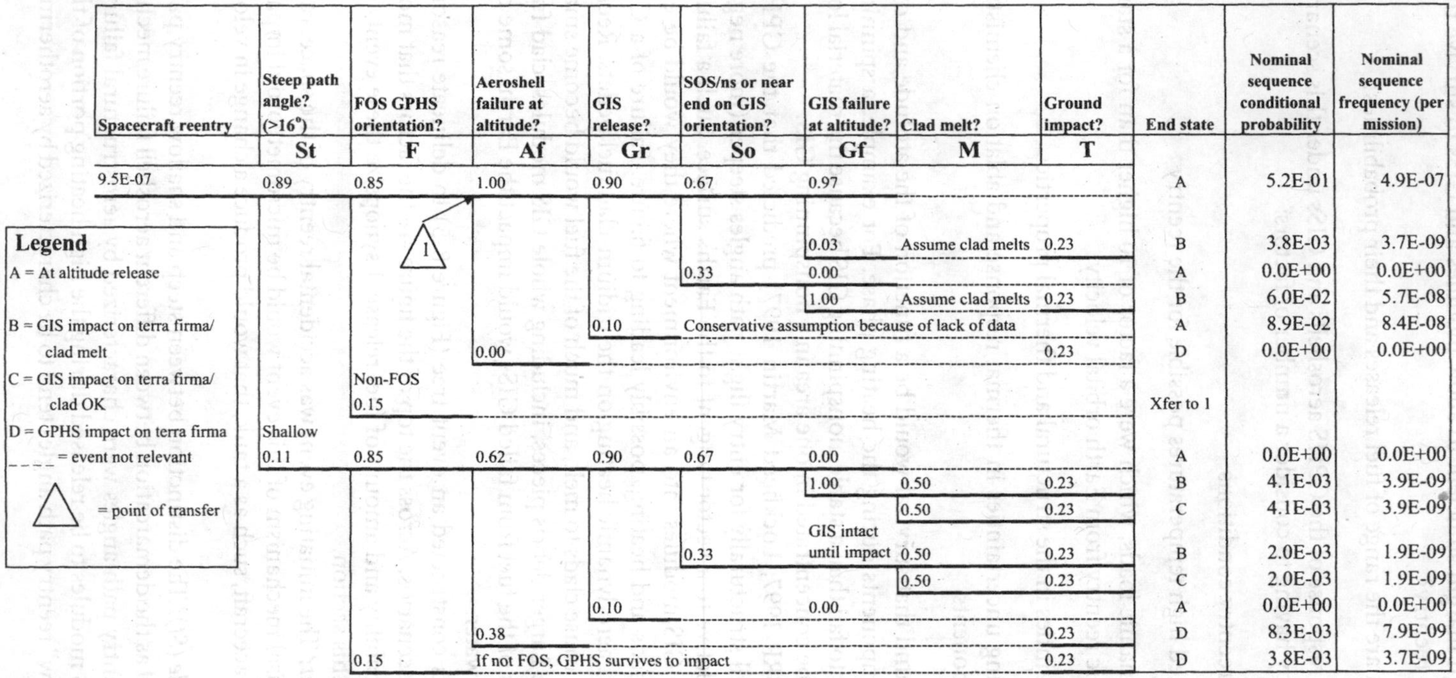

Spacecraft reentry	Steep path angle? (>16°) St	FOS GPHS orientation? F	Aeroshell failure at altitude? Af	GIS release? Gr	SOS/ns or near end on GIS orientation? So	GIS failure at altitude? Gf	Clad melt? M	Ground impact? T	End state	Nominal sequence conditional probability	Nominal sequence frequency (per mission)
9.5E-07	0.89	0.85	1.00	0.90	0.67	0.97			A	5.2E-01	4.9E-07
						0.03	Assume clad melts	0.23	B	3.8E-03	3.7E-09
					0.33	0.00			A	0.0E+00	0.0E+00
						1.00	Assume clad melts	0.23	B	6.0E-02	5.7E-08
				0.10	Conservative assumption because of lack of data				A	8.9E-02	8.4E-08
			0.00					0.23	D	0.0E+00	0.0E+00
		Non-FOS 0.15							Xfer to 1		
	Shallow 0.11	0.85	0.62	0.90	0.67	0.00			A	0.0E+00	0.0E+00
						1.00	0.50	0.23	B	4.1E-03	3.9E-09
							0.50	0.23	C	4.1E-03	3.9E-09
					0.33	GIS intact until impact	0.50	0.23	B	2.0E-03	1.9E-09
							0.50	0.23	C	2.0E-03	1.9E-09
				0.10		0.00			A	0.0E+00	0.0E+00
			0.38					0.23	D	8.3E-03	7.9E-09
		0.15	If not FOS, GPHS survives to impact					0.23	D	3.8E-03	3.7E-09

Legend

A = At altitude release
B = GIS impact on terra firma/ clad melt
C = GIS impact on terra firma/ clad OK
D = GPHS impact on terra firma
- - - - = event not relevant
△ = point of transfer

Figure 8-5. ***Earth-Gravity-Assist (EGA) Accident Event Tree***

Notes: FOS = face-on-stable; GPHS = general-purpose heat-source (modules); GIS = graphite-impact shell; SOS/ns = side-on-stable /not spinning; St = steep path angle; F = GPHS orientation; Af = aeroshell failure at altitude; Gr = GIS release; So = near end-on GIS orientation; Gf = at-altitude GIS failure; M = clad melt; T = terra firma impact.

Face-on-stable (FOS) GPHS orientation (F)? Reentering spacecraft can exhibit a variety of motions. Rotational or tumbling motion, however, would tend to be highly damped early in the trajectory, achieving a few revolutions per minute before releasing the aeroshells. Wind-tunnel data indicated a weakly statically stable trim point at FOS orientation (INSRP 1997b).

Aeroshell failure at altitude (Af)? Calculations suggested that aerostructural failures would occur early in the heat pulse, before peak thermal loading and temperature, for reentry path angles of approximately −16° and steeper (Baker and Nelson 1998). Therefore, there was essentially no uncertainty that structural failure would occur for both FOS and non-FOS orientations during steep reentry. For shallow reentry angles, however, aeroshell failure was highly dependent on modeling assumptions. INSRP mathematically included the effect of these alternative models in the uncertainty analysis that led to the probability estimates shown in Figure 8-5.

GIS release (Gr)? Data from ballistic reentry vehicles indicated that aeroshell thermostructural failure would likely lead to virtual disintegration of the aeroshell with little chance for GIS retention. Aeroshell breakup, then, would lead to free-falling GIS modules as independent ballistic bodies.

Side-on-stable (SOS)/not spinning (ns) or near end-on GIS orientation (So)? Wind-tunnel data and *Cosmos* data (Hobbs 1973; Hanafee 1978) indicated that GIS modules would exhibit two stable trim points with a preference for SOS orientation.

At-altitude GIS failure (Gf)? Thermal ablation appeared to be the only reasonable failure mode for the GIS modules. Similar to aeroshells, failure might or might not be calculated to occur depending on the modeling assumptions. INSRP included modeling uncertainties in developing the probability estimates. I discuss this in more detail later in this chapter.

Clad melt (M)? Fuel cladding was calculated to melt if the GIS modules were exposed to the airstream following aeroshell failure.

Terra firma impact (T)? GIS and aeroshells may not be effective at preventing additional fuel release on ground impact. INSRP had previously determined that potential releases of fuel into large bodies of water were considered to be effectively isolated from people (INSRP 1990).

Each of these events was associated with a probability of occurrence. The numbers shown in Figure 8-5 are means of the underlying probability distributions for the event. The probability of each scenario, therefore, is a product of the branch point probabilities and the initiating event probability. To obtain the total end state probability, analysts summed the probabilities leading to the same end state. For example, the total estimated probability of at-altitude releases is the sum of the probabilities of scenarios that end in end state A. The analysis indicated that the likelihood of an aeroshell surviving the reentry without releasing fuel would be relatively small. Event tree framework allowed the analysts to visualize alternative reentry scenarios and calculate the uncertainty in the probability of occurrence of each event.

8.8 Example Development of Risk Estimates Including Model Uncertainties

When developing the risk assessment, INSRP had to grapple with the variety of potential models used to account for phenomena such as heat transfer and ablation chemistry during the flight of modules through the atmosphere. Any results, therefore, had to characterize the uncertainty introduced by variations in models. An example is the calculation of the probability of GIS failure at altitude in which alternative hypotheses about heat transfer and ablation chemistry created many alternative models.

INSRP sought the probability of aeroshell or GIS failure resulting from burn-through as a function of reentry angle, *A*. This was particularly important when attempting to calculate GIS ablation because sensitivity studies that used deterministic calculations of ablation during reentry showed that GIS failure depended on the hypothesized model used in the calculation. Alternative models stem from many alternative data sets associated with (1) total heat transfer into and out of the GIS or aeroshell and (2) carbon ablation chemistry. A heat transfer factor, H_F, was defined to span the range of effects of alternative heat transfer models. The factor is multiplicative on the deterministic heat transfer calculational results and is normalized such that the "best estimate" is unity. Sensitivity studies allowed development of a probability density function (PDF) of the heat transfer factor, $p(H_F)$, which is presented in Figure 8-6.

Uncertainty in calculated graphite ablation arises from thermochemical properties, particularly the carbon species vapor pressure, and other properties that determine carbon species chemical enthalpies. In addition, carbon mechanical erosion is also a possible ablation enhancement mechanism. A "chemistry factor" (C_F) was defined such that the nominal best estimate calculation is represented by $C_F = 1$. Again, sensitivity studies, combined with alternative chemistry

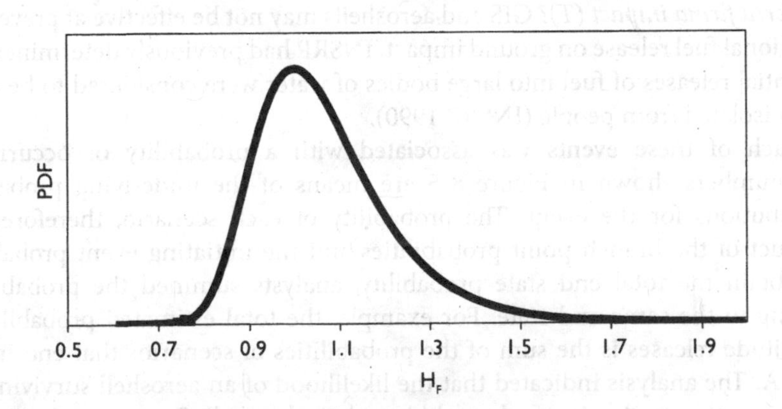

Figure 8-6. ***Representation of Heat Transfer Model Uncertainty***

Notes: PDF = probability density function; H_F = heat transfer factor.

models from the literature, allow development of a PDF of the chemistry factor, $p(C_F)$. The probability of GIS or aeroshell failure, therefore, can be expressed as $f_G(A_r\,|H_F, C_F)$, which is the conditional probability of GIS failure for reentry angle A_r given the joint occurrence of H_F and C_F.

For each A_r and H_F/C_F combination of interest, a stagnation point ablation calculation on the face of the GIS or aeroshell estimated whether the wall recession (i.e., ablation) was greater than or equal to the original wall thickness. In this way, a response surface for f_G was developed in which the elements were either one, if recession was greater than or equal to the original wall thickness, or zero, if not. For a particular GIS orientation, Equation 8-1 provides the relevant formula (Frank 2000).

$$f_G(Ar) \cong \iint f_G(A_r \mid H_F, C_F)\, p(H_F) p(C_F) dC_F dH_F \qquad (8\text{-}1)$$

The solution was performed numerically using discretized probability distributions for H_F and C_F across a discretized response surface. An example result is $f_G(-16° < A_r < -90°) = 0.97$ for the SOS/ns orientation. This analysis ascertained that there would be GIS failures during the EGA reentry accident with a high probability. In the next section, I present the risk model, represented as a set of equations, for estimating the amount of fuel released.

8.9 Example Source Term Calculations

Each end state in the event tree is associated with consequences. One of the consequences relevant to this risk assessment was the amount of fuel released.

Each reentry was characterized by a reentry angle, A_r, module orientation, O, and the probability of failure of an individual module, $P_M(A_r, O)$. The number of module failures in air, n_m, is a function of these and the total number of modules available. Module orientation and reentry angle were treated as random variables with probabilities P_O and P_A. A binomial distribution, $P_m(n_m \mid N_M, q_M)$, expressed the probability, P_m, of n_m module failures in air given the number of modules, N_M, and the probability of failure for each module, q_M. Use of the binomial is valid because each module failure is independent of all others. Thus, for a particular orientation and reentry angle, the analyst found that:

$$P_m(n_m \mid N_M, q_M) = \frac{N_M!}{n_m!(N_M - n_m!)}\, q_M^{\,n_m} (1 - q_M)^{N_M - n_m} \qquad (8\text{-}2)$$

The mean number of failed modules would be $q_M N_M$. In Equation 8-2, $q_M = (P_A)(P_O)P_M(A_r O)$, which is the probability of a single module failure for a specified reentry angle and orientation. N_M is the total number of binomial trials that corresponds to the total number of modules (54) on the *Cassini* spacecraft's three RTGs.

Because GIS are inside of modules, the probability of n_g GIS failures in air, $P_G(n_g)$, depends on the previously calculated number of module failures and is binomially distributed as follows:

$$P_g(n_g \mid N_G, q_G) = \frac{N_G!}{n_g!(N_G - n_g!)} q_G^{\,n_g}(1 - q_G)^{N_G - n_g} \tag{8-3}$$

The mean number of failed GISs would be $N_G q_G$. In Equation 8-3, $N_G = N_{GM}(n_m)$and $q_G = (P_{GO})(P_{GR})f_G$. N_{GM} is the number of GISs in a module ($N_{GM} = 2$); n_m is given by Equation 8-2; P_{GO} is the probability of a GIS orientation (e.g., probability of an SOS/ns orientation); P_{GR} is the probability of GIS release from the module given a module failure; and f_G is the probability of an individual GIS failure for a particular reentry angle and orientation as given by Equation 8-1. P_{GO} has a mean of 0.67 for SOS/ns orientation. Similarly, the mean of P_{GR} is 0.9. The quantities P_{GO}, P_{GR}, and f_G were relevant to both the calculation of the scenario probabilities and in the source term estimation.

Figure 8-7 presents results for the total amount of fuel released into the atmosphere. The separation of the curves along the ordinate (y-axis) derives primarily from the uncertainty in scenario frequency. The variation in results along the

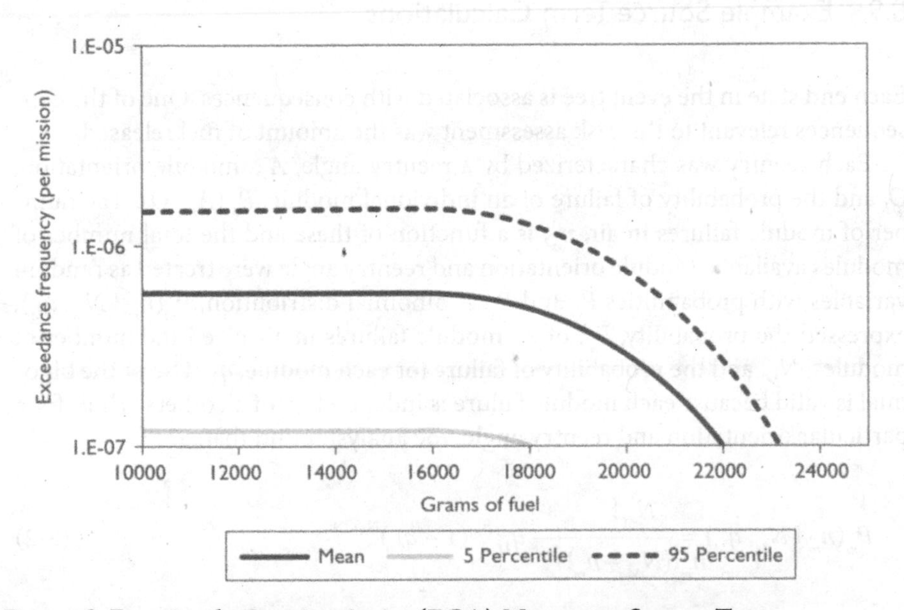

Figure 8-7. ***Earth-Gravity-Assist (EGA) Maneuver Source Term***

Note: Interagency Nuclear Safety Review Panel (INSRP 1997) estimate in grams of plutonium dioxide (PuO_2).

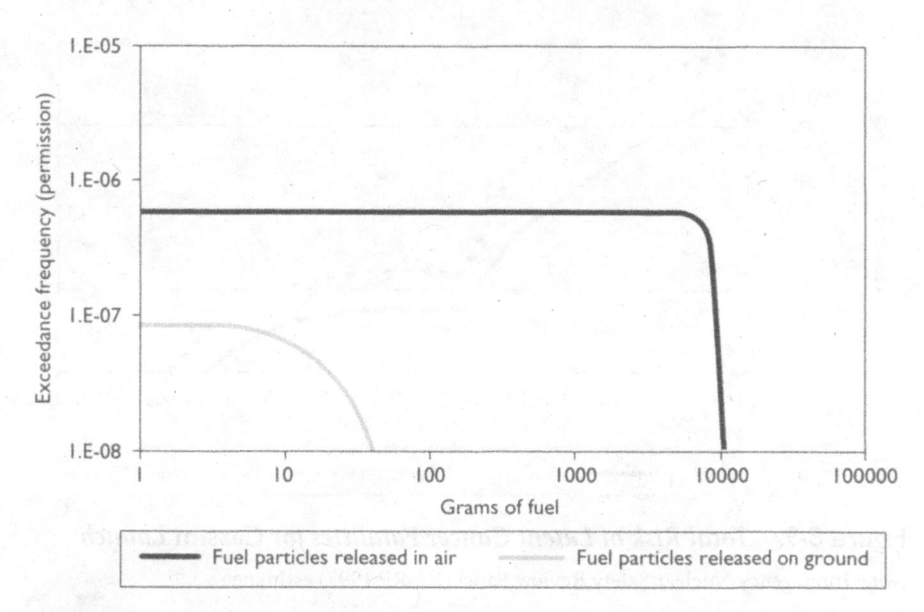

Figure 8-8. ***Comparison of Airborne versus Ground-Deposited Respirable Plutonium Dioxide (PuO_2) Fuel***

Note: Interagency Nuclear Safety Review Panel (INSRP 1997) estimate in grams of fuel.

abscissa (*x*-axis) derives from the source term parameters and models as described previously. Figure 8-8 compares the amount of respirable fuel (i.e., less than 10 μm) released into the atmosphere and initially suspended in air with the respirable fuel that results from the initial deposit of fuel on the ground.

8.10 The Decision Process: Launch or Don't Launch?

The Executive Office of the President of the United States was interested in the bottom-line consequences of accidents, which in this case was the potential number of latent cancer fatalities. Historically, the risk of radioactivity release from U.S. launches is very low, and the DM wanted to carry that historical record forward with the *Cassini* mission. The Executive Office was particularly interested in the total from all accident scenarios and the subtotal from the EGA reentry accident, because the latter represented the highest potential consequences resulting from a *Cassini* accident.

The total number of latent cancer fatalities associated with PuO_2 released in any of the three accident categories was estimated over a 50-year period. Because of the large dilution of releases in air, there was a distinct likelihood of no health effects from this accident. However, the traditional way of conservatively estimating the statistical probability of health effects is to hypothesize linearity

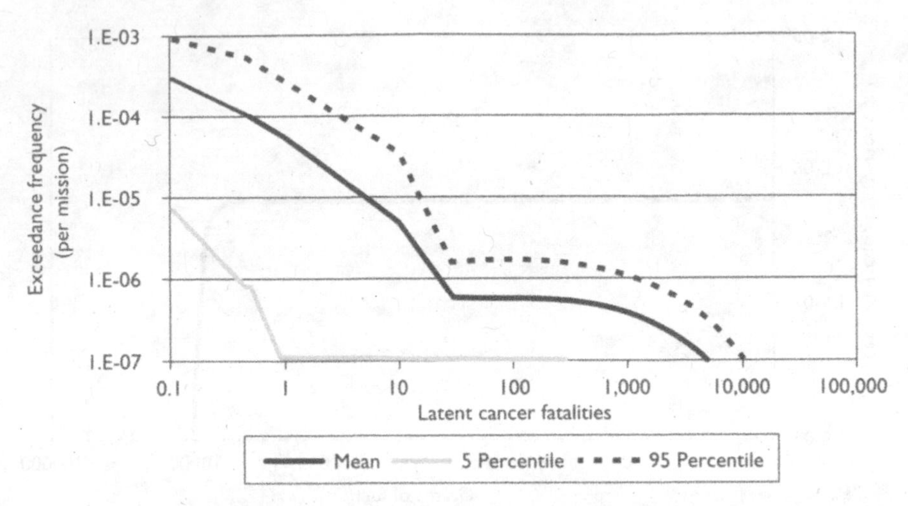

Figure 8-9. ***Total Risk of Latent Cancer Fatalities for* Cassini *Launch***
Note: Interagency Nuclear Safety Review Panel (INSRP 1997) estimate.

in the dose to health effects relationship for low doses. Making this hypothesis and distributing the source term uniformly around the world, the source term (in curies) can be converted, using simple conversion factors, to health effects in terms of statistical occurrences of latent cancers over a 50-year period (INSRP 1997). These estimates are subject to uncertainty associated with variations in the population density, as well as variations with respect to intake factors, pathway factors, and human susceptibility within a population. The total risks of latent cancer fatalities of all three accident categories were summed, and the INSRP-generated results are shown in Figure 8-9. Notice that the analysis yielded a mean probability of one or more latent cancer fatalities of approximately 1 in 14,000 missions.

When the analysis was complete, both the *Cassini* program and INSRP were invited to make presentations to OSTP about their method, results, and conclusions. The OSTP retained two additional consultants to help them understand the two approaches, explain the differences, and assess the credibility of the analyses. Both studies found the risk to be low. OSTP was also interested in the following four areas:

- The historical failure rate of 1 in 10 per mission of the *Titan IV/Centaur* vehicle even though calculated amounts of fuel release and number of cancer fatalities from launch accidents were low. The concern was about the reaction of the Florida residents to an explosion of the launch vehicle even if no radioactivity was released.

- The perception of radioactivity exposure of local Florida residents to plutonium, no matter how small it was calculated to be. Public perception of nuclear risks differs from calculated estimates of risk made by experts.
- The perception of the reentry of any part of the spacecraft because the international public may perceive a global risk even if no radioactivity was released.
- The perception of the robustness of the risk assessment process, including the perceived discrepancies between the FSAR and the SER, particularly differences between the INSRP and FSAR source term estimates associated with the EGA reentry accidents.

The last item was the key to assuaging concerns about the other three. A risk assessment that was peer reviewed and found to be scientifically defensible and comprehensive, and that showed small estimated probabilities of fuel release and latent cancer fatalities, was sufficient reason to grant approval to launch.

This decisionmaking process was focused on the *Cassini* mission itself. Future missions were not of concern to the DM, and experience from past missions was irrelevant to the DM's ruminations. The DM was not concerned with ways to improve the risk-assessment technology and database.

On the strength of the INSRP and FSAR analyses, combined with the recommendations of the consultants, the assistant to the president for science and technology recommended approval for the nuclear materials to be launched. The mission was a success.

Notes

1. In this chapter, I use the word *module* to indicate a GPHS module consisting of one aeroshell, two graphite-impact shells, and four iridium clads with associated structural components.
2. I use the term *fuel* in this chapter to indicate the heat-generating plutonium-based mixture that resides in the iridium clads within the RTGs.
3. A planetary gravity assist to a spacecraft occurs when its trajectory brings it close to a planet such that the planet's gravity accelerates it by "pulling" it faster as the spacecraft hurtles past.

References

Baker, R., and D. Nelson. 1998. Radioisotope Thermal Generator Survivability for Earth Reentries. *Proceedings of the International Conference on Probabilistic Safety and Management (PSAM4).* New York: Springer-Verlag.

Frank, M.V. 2000. Probabilistic Analysis of the Inadvertent Reentry of the *Cassini* Spacecraft's Radioisotope Thermoelectric Generators. *Risk Assessment* 20(2): 251–260.

Hanafee, J.E. 1978. *Analysis of Beryllium Parts for Cosmos 954.* UCRL-52597. Livermore, CA: Lawrence Livermore Laboratory.

Hobbs, R.B. 1973. MHW-HSA Aerodynamic Performance Test Report. Data memo 9151-001. Valley Forge, PA: General Electric Company, Re-Entry Systems Department, Aerothermodynamics Engineering Laboratory.

INSRP (Interagency Nuclear Safety Review Panel). 1990. *Biomedical and Environmental Effects Working Group Report for* Ulysses. INSRP 90-06. Washington, DC: INSRP.

———. 1997. *Supporting Technical Studies Prepared for the* Cassini *Mission Interagency Nuclear Safety Review Panel Safety Evaluation Report.*

JPL (Jet Propulsion Laboratory). 1993. Cassini *Program Environmental Impact Statement Supporting Study. Vol. 3:* Cassini *Earth Swingby Plan.* JPL D-10178-3. Pasadena, CA: JPL.

———. 1997. Cassini *Earth Swingby Plan Supplement.* JPL D-10178-3. Pasadena, CA: JPL.

Lockheed Martin. 1997. *General Purpose Heat Source-Radioisotope Thermoelectric Generators in Support of the* Cassini *Mission Final Safety Analysis Report (FSAR).* CDRL C.3. Valley Forge, PA: Lockheed Martin Missiles and Space Company.

NASA. 1995. *Final Environmental Impact Statement for the* Cassini *Mission.* Washington, DC: NASA.

———. 1997. *Final Supplemental Environmental Impact Statement for the* Cassini *Mission.* Washington, DC: NASA.

NSAM (National Security Action Memorandums). 1963. *Large-Scale Scientific or Technological Experiments with Possible Adverse Environmental Effects.* NSAM 235, April 17. Available online at http://www.fas.org/irp/offdocs/nsam-jfk/nsam-235.htm (accessed June 7, 2007).

———. 1961. NSAM 50, May 12, 1961. Available online at http://www.jfklibrary.org/Historical+Resources/Archives/Reference+Desk/NSAMs.htm (accessed June 10, 2007)

NSC (National Security Council). 1977. *Scientific or Technological Experiments with Possible Large-Scale Adverse Environmental Effects and Launch of Nuclear Systems into Space.* Presidential Directive. PD/NSC-25. December 14. Available online at http://www.fas.org/irp/offdocs/pd/pd25.pdf (accessed June 10, 2007). Updated by *Revision to PD/NSC-25, dated December 14, 1977, entitled Scientific or Technological Experiments with Possible Large-Scale Adverse Environmental Effects and Launch of Nuclear Systems into Space,* May 8, 1996.

CHAPTER 9

Mars Micro-Meteorology (Micro-Met) Stations

IN 1993, U.S., EUROPEAN, AND RUSSIAN space agencies worked together to create the International Mars Exploration Working Group. The group's charter was to define a common strategy for robotic and human exploration of Mars. One of the working group's recommendations was that a network of small stations be installed on the surface of Mars. These stations would be designed to gather the data necessary to understand the planet's internal structure, shape and rotational state, magnetic properties, and atmospheric circulation and weather patterns, among other factors.

The near-term contribution of the United States toward Mars exploration (more than a decade ago) was the Mars Environmental Survey program, conducted under the auspices of NASA's Ames Research Center. Its goal was to establish a dozen (or more) small robotics stations on Mars during the twenty-first century to study geology, surface chemistry, and meteorology. As part of this program, the Ames team developed a conceptual design of the Mars Micro-Meteorology (Micro-Met) Station mission in the 1990s for a possible flight to Mars in the first couple of decades of this century. The mission's objective was to establish a network of miniature meteorology stations on the surface of Mars.

In this case, the decisionmaker (DM) was the mission architecture project engineer. He identified an opportunity to use risk-assessment techniques to help decide on key aspects of the architecture, specifically the number of Micro-Met stations that should be deployed in order to give reasonable assurance that 12 stations would operate for a Martian year.

9.1 The Decision Opportunity

As pictured in Figure 9-1, the concept was to deploy at least 12 small stations (each weighing approximately 3 kg [6.6 lb]) by centrifugal propulsion from a spacecraft that would approach Mars on a fly-by trajectory. In essence, each of these miniature meteorological stations would be housed within a protective

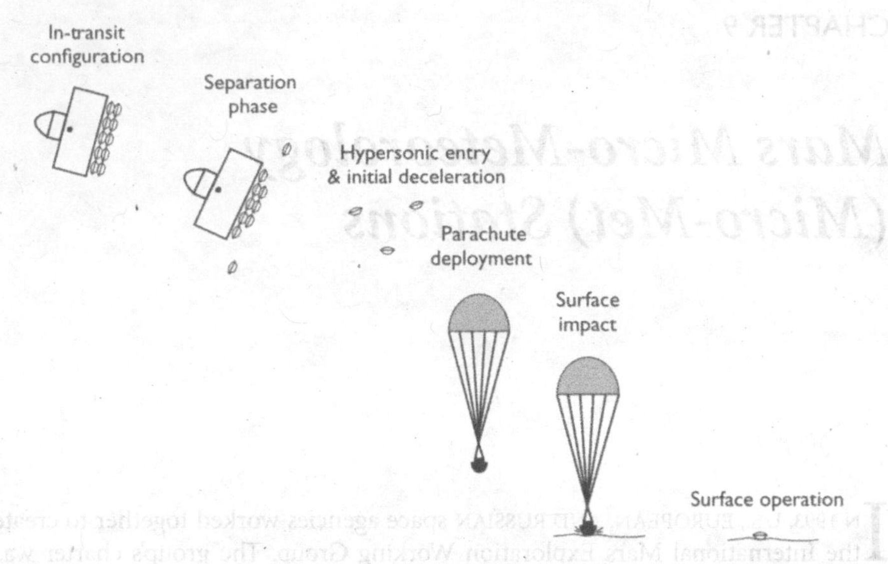

Figure 9-1. ***Conceptualization of Approach to Mars***

landing device and catapulted toward the Mars surface. The landing vehicle would consist of a protective aeroshell (shaped somewhat like an oyster or clamshell), insulation and cushion material, a parachute, pyrotechnic devices, and a Micro-Met weather station. The landing vehicle would enter the Mars atmosphere at approximately 7 km/sec (15,000 mph). Aerobraking would cause deceleration at the rate of about 20 g. At approximately 9 km (6 mi) above the surface, pyrotechnic devices would eject the top of the aeroshell and deploy a parachute. This would slow the Micro-Met station to freeway speed, about 30 m/sec (63 mph) just before impact. Impact deceleration would be on the order of 1,000 g.

After landing and successful startup, each miniature weather station would take pressure and temperature measurements periodically (about every Earth hour). With the successful deployment and operation of 12 stations, a meteorological mapping of Mars, including pressures, temperatures, and wind fields, could be achieved in one Martian year (687 Earth days). Each station would store about 30 days worth of measurements and transmit its data to an orbiter at 30-day intervals. The orbiter would, in turn, transmit the data back to Earth.

9.2 The Problem Statement

The DM gave the risk analysts three questions to answer. Given that the space vehicle could successfully catapult the stations toward Mars,

1. What is the predicted reliability of successfully completing a mission of at least 12 stations operating for one Martian year?
2. What changes (if any) to the mission architecture, the landing vehicle, and each Micro-Met station would be required to provide a 90% or greater chance of success?
3. Given the preliminary nature of the station architecture, what are the uncertainties in the results?

There were two constraints. First was the volume and mass available on launch vehicles. The *Delta* and the *Atlas* were two options for the launch vehicles. In both cases, the Micro-Met stations would be a piggyback payload. The mission would require a launch vehicle that had excess mass and volume capacity. Once identified, this relatively small spacecraft would be launched with the primary payload. One launch option had sufficient excess capacity for 15 Micro-Met stations, and the other might be able to accommodate 20 to 25 stations.

The second constraint was cost. Launch vehicle costs would be comparable and, with all else being equal, selecting the lowest cost alternative would be desirable, although not essential. The scientific goals of high reliability in data collection and transmittal were the most important to achieve, however. This study was done very early in order to develop a proposal to obtain funding for the project, as there was no project budget at the time. A proposal that was too expensive or that lacked high assurance of meeting the mission goals would simply not win approval to proceed.

9.3 The Objective and the Attributes

The attribute of interest to the DM was mission risk, or its complement, mission reliability. This assessment, then, was a single-attribute decision analysis. The DM's questions implied the need to perform a risk assessment. The scope was to perform a quantitative risk assessment to estimate mission unreliability with uncertainties as a function of the number of launched Micro-Met weather stations. The objective was to estimate the number of such stations needed to achieve a mean reliability of at least 90%.

9.4 The Alternatives

Each design with a different number of Micro-Met stations constituted an alternative. The analysis included designs with 12 through 25 stations.

9.5 The Risk Model

Development of the risk model proceeded from understanding of the operation and design concepts to development of fault trees and the associated mathematical model. The risk model development ended with evaluation of the model using readily available sources of equipment reliability information.

9.5.1 Micro-Met Mission Architecture, Configuration, and Assumptions

Figure 9-2 diagrams the sequence of functions of the mission. Wishing to limit the scope, the DM set forth an assumption that the spacecraft would be successfully actuated and checked out while still on board the carrier spacecraft (before being jettisoned toward Mars). Once away from the carrier spacecraft, pyrotechnic devices would thrust open the aft cover to allow the parachute, which would also be deployed by pyrotechnics, to eject. Yet another set of pyrotechnic devices would jettison the upper part of the aeroshell, exposing the Micro-Met station instrument package before landing. Each pyrotechnic device would be NASA standard issue. Synchronous firings of multiple pyrotechnic devices with

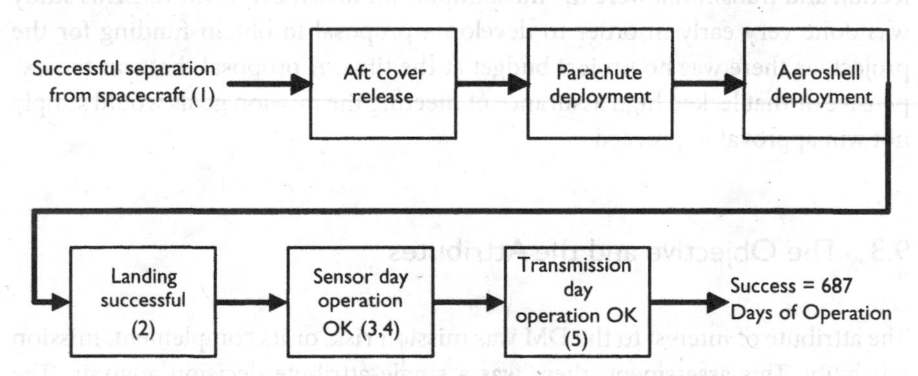

Figure 9-2. ***Mission Functional Sequence and Analysis Assumptions***

Notes: (1) Assumes controller, central processing unit, and clock are initialized by spacecraft and there are no latent failures resulting from launch, cruise, and separation.

(2) Assumes that structural failures of lander shells, insulation, and aerobrakes are small contributors to unreliability.

(3) Assumes that lightweight radioisotope heater units and thermoelectric converter are small contributors to unreliability.

(4) Includes all active components except those used only for transmission; components assumed to be used continuously for 687 Earth days.

(5) Includes switches, power controllers, receivers, and transmitters used only for transmission day operation; components assumed to be used every 30 days for a total intermittent duty of 23 days.

extremely precise timing would be required to deploy the aft cover and aeroshell. Data from the space shuttle program indicated that nonsynchronous firing might be an important failure mode.

A successful landing would simply mean that the instrument package had survived impact and had landed in a location that allowed a view of the sky. The station would be designed such that its antennae would have a view irrespective of its orientation on the Martian surface. Here, problems could be the instrument package being buried in dust or landing in permanent shadow, but assessing structural and landing location failures was not within the scope of this analysis. The notes to Figure 9-2 document other analysis assumptions. Figure 9-3 displays a block diagram of the Micro-Met station with some configuration details.

After landing, the Micro-Met station would begin its mission of collecting, storing, retrieving, and transmitting data. "Sensor-day" operations would use power devices, processor and memory with interface electronics, and sensor components. On "transmission days," the station would perform additional tasks, operating the transmitters and receivers (for handshake protocol) and the

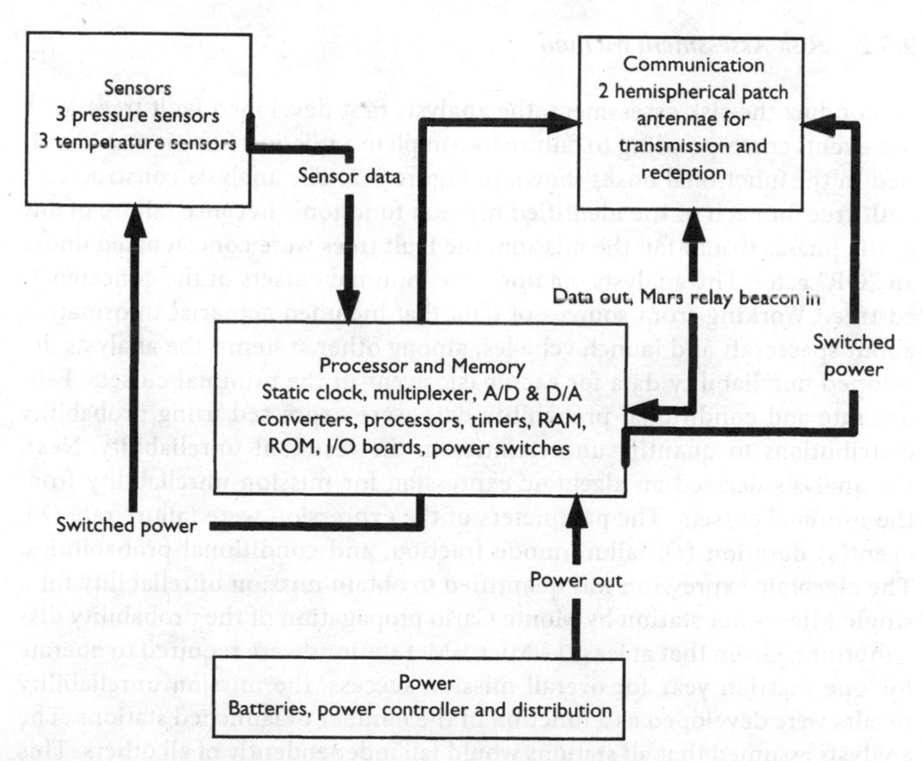

Figure 9-3 *Major Elements of the Micro-Met Station*

Notes: A/D = analog/digital; D/A = digital/analog; RAM = random-access memory; ROM = read-only memory; I/O = input/output.

associated switches and power controllers. The station would experience an increased power demand on transmission days.

Station features would include three barometric sensors with ranges from 0 to 50 mbar. Thermocouples for temperature measurements would be needed for calibration because of the temperature dependence of the barometer readings. These would have a range of 100K to 370K. Power would be supplied by one lightweight radioisotope heater unit (LWRHU), which would generate about 1 W of thermal power. The LWRHU would be coupled to a thermoelectric converter that would generate about 40 mW of steady-state power, along with two $LiTiS_2$ batteries that would produce 4,000 mW-h at 4 V. In the LWRHU, the source of power would be radioactive decay of plutonium 238 (Pu-238). This power source is extremely reliable, with no known failures to produce power in space. Another on-board LWRHU would provide enough heat to preserve the batteries during the Martian night.

A timer (clock) would control the on–off operations of components, triggering measurements 50 times every Martian sol.[1] Transmissions would be triggered by receiving a beacon from an orbiting communication relay satellite.

9.5.2 *Risk Assessment Method*

To conduct the risk assessment, the analysts first developed fault trees, with top events corresponding to failure to complete a mission function as identified in the functional boxes shown in Figure 9-2. The analysts constructed a fault tree for each of the identified mission functions. Because failure of any of the phases would fail the mission, the fault trees were concatenated under an "OR" gate.[2] The analysts obtained the minimal cutsets of the concatenated tree.[3] Working from sources of data that included actuarial information about spacecraft and launch vehicles, among other systems, the analysts developed unreliability data for each basic event in the minimal cutsets. Failure rate and conditional probability data were expressed using probability distributions to quantify uncertainties in the resultant unreliability. Next, the analysts derived an algebraic expression for mission unreliability from the minimal cutsets. The parameters of the expression were failure rate (λ), event(s) duration (t), failure mode fraction, and conditional probabilities. The algebraic expression was quantified to obtain mission unreliability for a single Micro-Met station by Monte Carlo propagation of the probability distributions. Given that at least 12 Micro-Met stations were required to operate for one Martian year for overall mission success, the mission unreliability results were developed as a function of the number of launched stations. The analysts assumed that all stations would fail independently of all others. This may not strictly be the case because the analysts could not exclude common cause failures, such as burial in dust, insufficient view of the sky, or design inadequacy, at that time. The overall mission success probability, P_S, for a given number of deployed stations, N_s, and an individual station reliability, r_s, would be

$$P_s(x \mid N_s, r_s) = \sum_{x=12}^{N_s} \frac{N_s!}{x!(N_s - x)!} r_s^x (1 - r_s)^{N_s - x} \tag{9-1}$$

where P_S is the cumulative distribution function for 12 or more (through N_s) deployed stations. The analysts calculated the individual station reliability, r_s, from the fault tree analysis.

9.5.3 *Data Development for Mission Risk Assessment*

Because this analysis was performed during conceptual design of a new mission and of new systems, the analysts used a variety of references for failure rates and probabilities. Two kinds of failure rates were sought: (1) failures per hour, which are indicative of continuously operating equipment; and (2) failures per demand or cycle, which are indicative of switching and cyclic equipment. For example, the pyrotechnic devices were characterized by failure per demand. Clocks and computers were characterized by failures per hour. The analysts also recognized that because this would be a new design, uncertainties would be larger than a risk assessment done for a mature, well-specified design.

The mission designers made a conscious effort to base component selection on either standard technology or technology of previous missions (such as *Viking* and *Mars Pathfinder*). They espoused a data development philosophy in which the components that would be used on the Micro-Met mission would be of the population found in the references. From this basic philosophy, it follows that the failure rates exhibited during the mission should be encompassed by the variability or range of failure rates derived from these references at similar environmental conditions. This included, for example, the use of failure rates associated with extreme environments to simulate Mars reentry environments. Components and failure rate information were cataloged in various references for each component of the model, and these exhibited a large variation in failure rates. Looking at each basic event in the fault tree, the analysts used regression analysis subject to a χ^2 goodness-of-fit test to analyze the variation of failure rates. They tested a variety of probability distributions, such as lognormal, Beta, exponential, normal, Rayleigh, and Weibull and used the best fit.

In a few instances, only a single failure rate could be found among the references. In those situations, the analysts treated the reference failure rate as the mean of a maximum entropy lognormal distribution. The concept of maximum entropy for defining a distribution has its derivation in information theory. According to Jaynes (1957), "It is the least biased estimate possible on the given information; i.e., it is maximally noncommittal with regard to missing information."

Table 9-1 presents an example of part of the database developed for this analysis.

Software and nonsynchronous pyrotechnic bolt firing required a somewhat different approach. The analysts estimated the software failure rate based on judgment and using two previous studies of digital controller software for flight systems (Dunn 1986, Dunn et al. 1994). Other studies found that nonsynchronous firing of pyrotechnic bolts during the descent phases could cause the bolts to jam and the aeroshell or aft hatch release function to fail. Space shuttle data indicated a similar problem with the solid rocket booster hold-down bolts during launch. These experience data were used as evidence for a likelihood function in a Bayesian analysis starting with a noninformative prior distribution. The analysts used the resulting posterior distribution as the failure probability for this analysis.

Table 9-1. *Example Data Entries for Micro-Met Risk Assessment*

Component name	*Failure modes*	*Component failure rate (units of 1E-06 per hour)*	
Pressure sensor	Incorrect reading during steady-state operation	5%	0.1
		Mean	3
		95%	10
Power switch	Fails to open or close	5%	0.05
		Mean	1
		95%	5
Power conditioner	Fails because of open circuit or short	5%	0.1
		Mean	0.7
		95%	2
Temperature sensor	Fails during power-up sequence	5%	0.1
		Mean	0.6
		95%	1.5
Battery	Low output	5%	0.5
		Mean	5
		95%	14
Power distribution	Fails to operate	5%	0.4
		Mean	5
		95%	17
Multiplexer	Fails to operate	5%	0.8
		Mean	3
		95%	6
Analog-to-digital converter	Fails to operate	5%	0.3
		Mean	3
		95%	10

9.5.4 *Risk Assessment Results*

Figure 9-4 shows the baseline results of the risk assessment. Mission unreliability is plotted as a function of the number of deployed stations under the assumption that 12 are required for successful data collection. The uncertainty in the unreliability is large and caused by the wide variability in failure rates of the components. The figure shows four curves—the 5th percentile, the 50th percentile, the mean, and the 95th percentile. If we draw a vertical line through the curves (e.g., at 17 total launch stations), we can interpret the locus of points of the intersection of the vertical line and the four curves as follows:

- There would be a 90% chance that the mission unreliability would be between 0.17 (at the 5th percentile) and nearly 1.0 (at the 95th percentile).
- There is an equal chance that the mission unreliability would be greater than or less than approximately 0.5.
- The mean unreliability accounting for the uncertainties would be approximately 0.65.

Despite the uncertainty, we can answer a fundamental question with high confidence by looking at Figure 9-4. Even at the lower confidence bound of

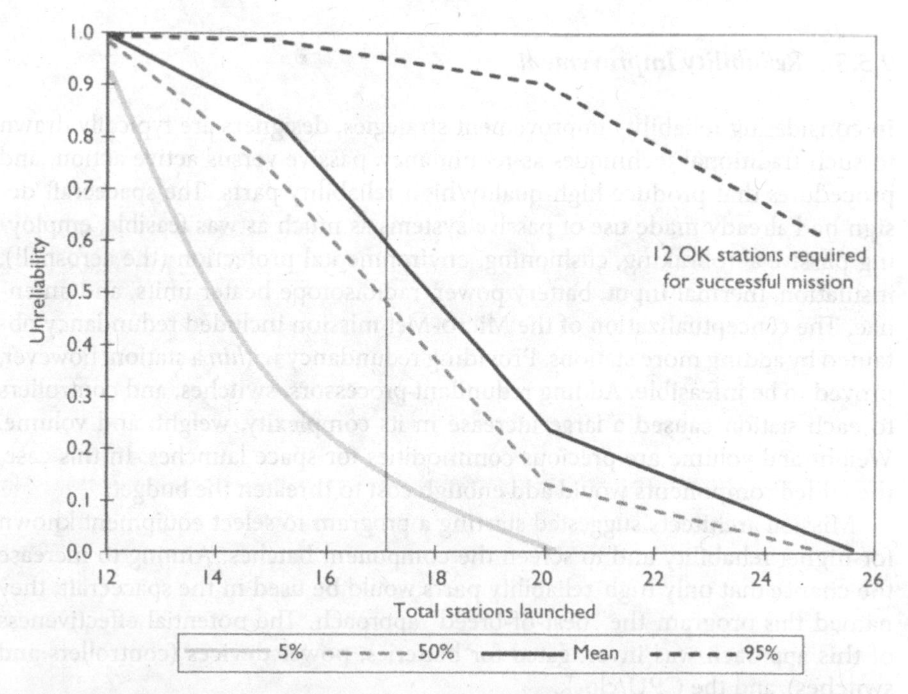

Figure 9-4. ***Mission Unreliability as a Function of Number of Stations Launched***

unreliability[4] (5th percentile), at least 19 Micro-Met stations would need to be deployed to achieve a 90% chance of overall mission success. Said another way, if all components behaved in a manner such that their cumulative effect was to produce the top 5% of reliability, at least 19 stations would need to be deployed to achieve 90% mission reliability (i.e., less than 10% mission unreliability). In the study, the analysts obtained a more robust decision about how many deployed stations would be required from the mean curve, which indicates that at least 24 stations would have to be deployed. Because this number was beyond the mass and volume capabilities of the proposed launch vehicles, it was unacceptable to the mission architects.

These results are predicated on the assumption that the components would act as if they are representative of the range of surrogate data found in the references. To investigate further, the analysts dissected the results to obtain the relative contributions of unreliability by component type. They found that five components contributed 90% of the individual Micro-Met station unreliability: batteries (29%), nonsynchronous pyrotechnic bolt firing (21%), power controllers (16%), power switches (14%), and processor (i.e., the central processing unit [CPU]) with clock (10%). The large contribution from power switches and controllers resulted from the large number of them in a series system design with no redundancy. The mission architects considered these areas for reliability improvement strategies.

9.5.5 *Reliability Improvement*

In considering reliability improvement strategies, designers are typically drawn to such traditional techniques as redundancy, passive versus active action, and procedures that produce high-quality/high-reliability parts. The spacecraft design had already made use of passive systems as much as was feasible, employing passive aerobraking, cushioning, environmental protection (the aeroshell), insulation, thermal input, battery power, radioisotope heater units, and antennae. The conceptualization of the Micro-Met mission included redundancy obtained by adding more stations. Providing redundancy *within* a station, however, proved to be infeasible. Adding redundant processors, switches, and controllers to each station caused a large increase in its complexity, weight, and volume. Weight and volume are precious commodities for space launches. In this case, the added components would add enough cost to threaten the budget.

Mission architects suggested starting a program to select equipment known for higher reliability and to screen the component batches. Aiming to increase the chance that only high-reliability parts would be used in the spacecraft, they named this program the "best-of-breed" approach. The potential effectiveness of this approach was investigated for batteries, power devices (controllers and switches), and the CPU/clock.

To guide the best-of-breed program, the risk analysts performed a sensitivity study to investigate the predicted effect of the best-of-breed approach. They assumed that the most unreliable 50% of the failure rate probability distribution could be eliminated. In other words, the analysis was performed again by using failure rate distributions that reflected the best 50% of the range found in the literature and used in the previous results. As Figure 9-5 shows, this produced a dramatic change in the predicted mission unreliability. In this figure, only the mean curves of the original base-case analysis and the best-of-breed analysis are compared. We can see that instead of the 24 stations for a 90% mission success probability, the best-of-breed strategy would require 19 stations.

Figure 9-6 is another presentation of the comparison between the base-case and the best-of-breed strategy. It expands the comparison to include the sensitivity of results to total launched landers (12, 15, 20, and 25), out of which 12 are required for success. Using the assumption that the best 25% of the equipment reliability could be obtained, the analysts performed another sensitivity study. This analysis showed an insignificant further reduction in the total mission reliability. This is a

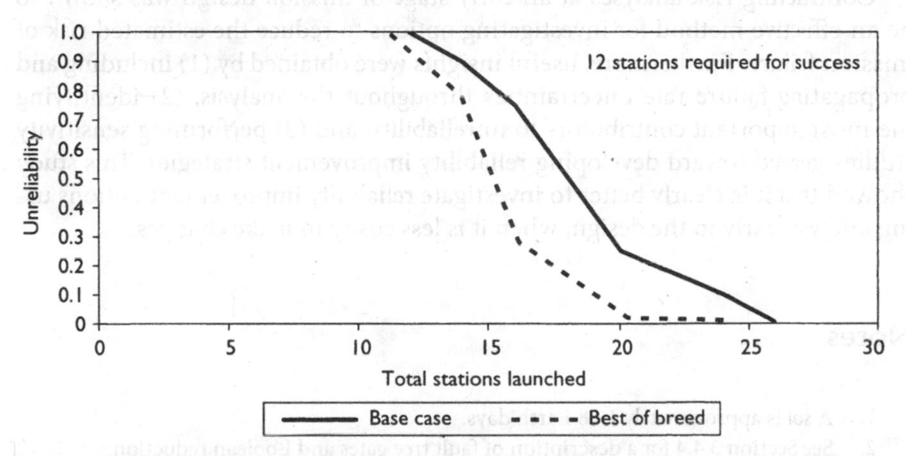

Figure 9-5. ***Mean of Base Case Compared to Mean of Best-of-Breed Strategy***

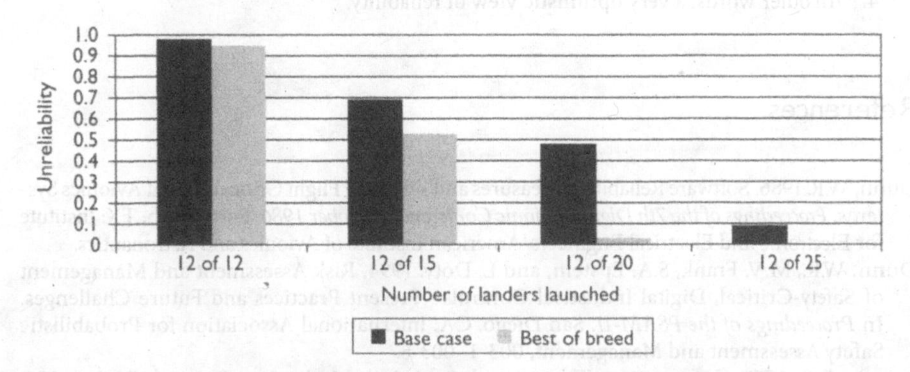

Figure 9-6. ***Incremental Base Case versus Best-of-Breed Comparison***

reasonable result because as the unreliability of the batteries, the power devices, and the CPU/clock are reduced, they reach the level of contribution of many other components. In other words, all other components of the Micro-Met design would contribute far more to unreliability than those for which the best-of-breed strategy was investigated. Any further reductions in mission unreliability, then, would have required changing the basic station design or implementing a best-of-breed strategy for all design components.

9.6 The DM's Reaction

The DM was surprised at the relatively low mission reliability and the number of stations needed to obtain an acceptably high reliability. But the analysis did show a feasible way to obtain the desired mission reliability goals.

Conducting risk analyses at an early stage of mission design was shown to be an effective method for investigating options to reduce the estimated risk of mission failure. Furthermore, useful insights were obtained by (1) including and propagating failure rate uncertainties throughout the analysis, (2) identifying the most important contributors to unreliability, and (3) performing sensitivity studies geared toward developing reliability improvement strategies. This study showed that it is clearly better to investigate reliability improvement options using analysis early in the design, when it is less costly to make changes.

Notes

1. A sol is approximately 1.88 Earth days.
2. See Section 3.4.4 for a description of fault tree gates and Boolean reduction.
3. Boolean reduction to minimal cutsets is important to avoid double counting component events used in more than one tree.
4. In other words, a very optimistic view of reliability.

References

Dunn, W.R. 1986. Software Reliability: Measures and Effects in Flight Critical Digital Avionics Systems. *Proceedings of the 7th Digital Avionic Conference, October 1986.* Fort Worth, TX: Institute for Electronic and Electrical Engineers/American Institute of Avionics and Astronautics.

Dunn, W.R., M.V. Frank, S.A. Epstein, and L. Doty. 1994. Risk Assessment and Management of Safety-Critical, Digital Industrial Controls—Present Practices and Future Challenges. In *Proceedings of the PSAM-II.* San Diego, CA: International Association for Probabilistic Safety Assessment and Management, 003-1–003-6.

Jaynes, E.T. 1957. Information Theory and Statistical Mechanics. *Physical Review* 106: 620–630.

CHAPTER 10

Choosing the Best Severe Accident Management Strategy

ACTIONS TAKEN TO REDUCE THE CONSEQUENCES of a severe accident either before it occurs or while it's happening are combined under the term "severe accident management." So what's the best strategy for managing a severe accident? The answer depends on your perspective. If we ask different individuals or groups (sometimes called "stakeholders") who have an interest in a decision about alternative safety strategies, each individual or group may make a different choice. This case study looks into choosing the best strategy that nuclear power plant operators can implement for coping with a particular nuclear reactor severe accident. A decision analysis, using probabilistic risk assessment (PRA) input, is developed from three perspectives: costs associated with radiological deaths; costs associated with loss of nuclear plant investment, repair, decommissioning, and litigation; and costs associated with regional financial losses resulting from a nuclear accident. We could argue that each of these is associated with somewhat different but not mutually exclusive stakeholder groups. For example, members of the public and the U.S. Nuclear Regulatory Commission (NRC) would be stakeholders because the NRC has a legislative mandate to ensure that nuclear power plants protect the public health and safety (NRC 2002). Federal regulations contained within 10 CFR 50 (NRC 2006a) and supporting regulatory guides (NRC 2006b) have been developed toward that goal. Regional business owners, state and local governments, and members of the public would be considered stakeholders if a nuclear accident affects regional economic activity and threatens public health. Because of their investment loss, nuclear power plant owners and operators would be stakeholders for a core melt accident, even if it releases no radiation and does not harm the public. We can also consider them to be part of the other two stakeholder groups because they are regional business owners and members of the public.

Consider the question: What does the phrase "the public health and safety" mean? In 1988, the NRC set qualitative and quantitative risk goals for nuclear reactors to clarify what it considers an acceptable level of public protection from nuclear power plant operation (see sidebar).

Acceptable Level of Public Protection against Nuclear Power Plant Operation: NRC Goals

Individual members of the public should be provided a level of protection from the consequences of nuclear power plant operation such that individuals bear no significant additional risk to life and health.

Societal risks to life and health from nuclear power plant operation should be comparable to or less than the risks due to electric generation by competing technologies and should not be a significant addition to other societal risks.

The risk to the average individual in the vicinity of a nuclear power plant of prompt fatalities that might result from reactor accidents should not exceed one-tenth of one percent (0.1%) of the sum of prompt fatality risks resulting from other accidents to which members of the U.S. population are generally exposed.

The risk to the population in the area near a nuclear power plant of cancer fatalities that might result from nuclear power plant operation should not exceed one-tenth of one percent (0.1%) of the sum of cancer fatality risks resulting from all other causes (NRC 1988).

Why did the NRC set forth these goals? The accident at Three Mile Island Unit 2 (near Middletown, Pennsylvania, March 28, 1979) caused extensive reexamination of how to ensure the public's health and safety. The NRC concluded that severe accidents constitute the major risk to the public associated with radioactive releases from nuclear power plant accidents. In 1985 the NRC published the "Policy Statement on Severe Reactor Accidents Regarding Future Designs and Existing Plants" in the *Federal Register* (NRC 1985). In this statement, the agency defined severe nuclear accidents as those in which the reactor core suffers substantial damage, whether or not there are serious off-site radiological consequences. The accident at Three Mile Island did not produce measurable off-site radiological consequences. The 1985 policy also stated that "existing plants pose no undue risk to public health and safety." The NRC, along with the nuclear power industry, sponsored research programs to investigate severe accident management, accompanying the development and issuance of the policy statement. Here are a few examples of the products of this research:

- PRAs of nuclear power plants that emphasized detailed investigation and analysis of the physical, chemical, and biological processes associated with severe nuclear accidents (e.g., NUREG-1150 [NRC 1990])
- PRAs, called individual plant evaluations (IPEs) and individual plant evaluations of external events (IPEEEs), of each large commercial nuclear power plant under 10 CFR 50 license (NRC 1989, 1991)

- Processes for development and assessment of strategies for coping with severe accidents (e.g., NUREG/CR-5543 [Hanson et al. 1991] and NUREG/CR-6056 [Kastenberg 1993])
- Guidance for nuclear power plant owners and operators on how to bring to closure all activities associated with severe accident issues (e.g., EPRI 1992).

10.1 The Decision Opportunity

During the late 1980s and early 1990s, working under a grant from the NRC, a group of university researchers developed a framework for choosing among alternative severe accident management strategies. These researchers were familiar with the PRA methods I discussed in Chapters 2 and 3. They were also familiar with proposed methods for selecting accident management strategies on the basis of criteria related solely to safety risk. For example, Hanson et al. (1991) proposed ranking strategies based on five criteria:

1. Likelihood that the nuclear reactor operators can successfully implement the strategy during a severe accident
2. Effectiveness of a strategy with respect to items such as reducing the probability of core damage, reducing the probability and amount of radionuclides released to the public, and the number of accident scenarios that the strategy can control
3. Minimization of negative effects (an example of a negative effect would be a partial release of pressure from the containment building, with accompanying radionuclide release, to delay containment failure)
4. Maximization of information readily available to operators
5. Maximization of procedures and equipment with which operators are familiar.

Three of the five criteria are related to ease of implementation by the nuclear power plant operators. The remaining two are related to reducing the safety risk of nuclear power plant accident scenarios, clearly a desire that everyone shares.

The researchers realized that choosing among alternative strategies, whether associated with severe accidents or not, is essentially a decision problem.

10.2 The Problem Statement

The researchers wanted to investigate how best to choose among severe accident management strategies from a broad and rigorously quantitative perspective.

For example, how would strategies compare with each other if both cost-effectiveness and safety risk were considered? And how would strategies compare from the perspectives of different stakeholders, such as the public and regional business leaders?

As I've discussed throughout this book, PRA is used to model accident scenarios. Recall that one of the most commonly used PRA techniques is an event tree, which is similar to a decision tree but is limited to chance nodes. It was a natural choice, therefore, for the researchers to decide to use a decision theoretic approach as a framework for their investigation.

The initial goal stemming from the problem statement was to develop a method for choosing among alternative severe accident management strategies. Accident management strategies, however, depend on the particular severe accident scenario—no single strategy fits all accident scenarios. Based on the results of NUREG-1150 (NRC 1990), the researchers selected a hypothetical severe accident scenario to use as an example to illustrate the method. The selected accident would begin with a loss of all electric power to a pressurized water reactor type of nuclear power plant. This accident scenario is sometimes called station blackout (SBO).[1]

10.3 The Objectives and the Attributes

The researchers wanted to develop a methodology that would be quantitative, would consider cost and safety, and would be flexible enough to represent different stakeholders. As a result, they set forth the following objectives and attributes:

- Develop a decision theoretic method to choose among alternative severe accident management strategies that quantitatively includes cost and safety attributes and applies to different stakeholders.
- Demonstrate the method using severe accident management strategies applicable to the SBO.

10.4 The Alternatives

Severe accident management strategies are derived from studying the sequence of events of the accident scenario being considered. In this case, researchers worked to gain a thorough understanding of the SBO sequence of events. I summarize this sequence of events in the paragraphs that follow.

Nuclear power plants derive electric power from the normal electric distribution system of the region. This is called off-site power. On-site power, which is produced by diesel generators and station batteries, is used to power key safety systems such as the emergency core cooling system, the power-operated relief valves, and the auxiliary feedwater system when off-site power is not available.

As depicted schematically in Figure 10-1, the nuclear reaction occurs in the nuclear fuel, which is contained within a reactor vessel with the control rods. The reactor vessel and the inner part of the steam generator are part of the first loop. There are typically three or four steam generators, but only one is shown in Figure 10-1. The first loop is housed inside a very strong containment building. Five major barriers are in place to prevent release of radionuclides to the public. First is the fuel itself, which is constructed to hold within it many of the radionuclides produced by nuclear fission. Second is the metal alloy cladding surrounding the nuclear fuel, which requires constant cooling to remove the heat the fuel generates. Third is the reactor vessel and the first loop, which is a sealed steel and metal alloy boundary. In the absence of one or more failures, the water and radioactive material in the first loop stays in the first loop. Fourth is the containment building, and fifth is the exclusion area, where members of the public are not allowed without permission.

The SBO scenario starts with loss of off-site power. The control rods automatically insert to stop the nuclear reaction when off-site power is lost. At the same time, the on-site emergency diesel generators attempt to start but fail to do so. These events cause a loss of the main source of water into the reactor's steam generators (which are similar to boilers) because the pumps that provide water

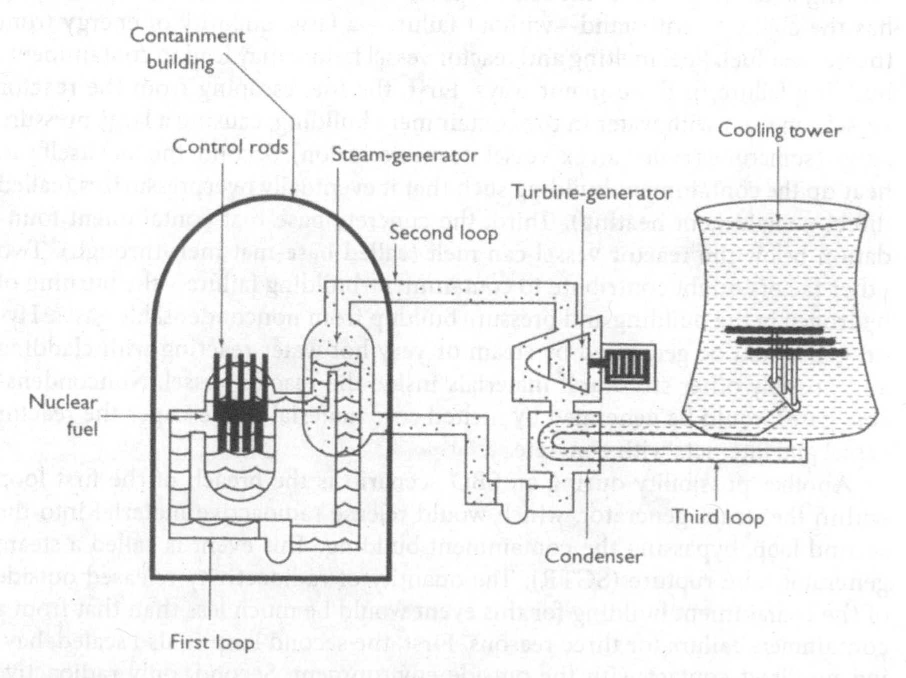

Figure 10-1. *Pressurized Water Reactor Schematic*

require off-site power to operate. Although there are three auxiliary sources of water for the steam generators, two of them would not work during this scenario because they require on-site diesel power. This SBO scenario further postulates that the third source, which requires battery power instead of diesel or off-site power, also fails. Without water flowing into the steam generators, the water they contain would typically boil away in tens of minutes.

Although the nuclear reaction has been shut down, the radionuclides in the nuclear fuel continue to create heat because heat generation is a by-product of radioactivity. This heat must be removed to prevent the fuel from overheating, becoming damaged, and possibly melting. A nuclear power plant is equipped with multiple safety systems that can cool the nuclear fuel. The SBO accident, however, prevents these from working because neither on-site nor off-site sources of electric power are available. If water into the steam generators is unavailable and the safety systems cannot operate for a sufficiently long time, the nuclear fuel begins to melt. This would breach the first barrier. A longer time is needed to breach the second barrier. This gives operators more time to restore electric power and begin water flow from the safety systems to prevent failure of the reactor vessel. If electric power is restored before the fuel melts, the accident scenario is terminated and the reactor is in a safe state.

If water flows into the reactor vessel, however, after the fuel begins to melt, there's still a small chance that the vessel will fail anyway. The fuel may have reconfigured to a shape or geometry that cannot be effectively cooled even with flowing water. In this case, the fourth barrier remains. The containment building has the ability to withstand—without failure—a large quantity of energy from the nuclear fuel. Fuel melting and reactor vessel failure may lead to containment-building failure in three major ways. First, the fuel escaping from the reactor vessel can react with water in the containment building, causing a large pressure surge (sometimes called an ex-vessel steam explosion). Second, the fuel itself can heat up the containment building such that it eventually overpressurizes (called direct containment heating). Third, the concrete base-mat containment foundation below the reactor vessel can melt (called base-mat melt-through). Two other factors might contribute to containment-building failure—the burning of hydrogen in the building and pressure buildup from noncondensable gases. Hydrogen would be generated by steam or very hot water reacting with cladding and possibly other structural materials inside the reactor vessel. Noncondensable gases would be generated by melted core material that escapes the reactor vessel and interacts with concrete.

Another possibility during an SBO scenario is the breach of the first loop within the steam generator, which would release radioactive material into the second loop, bypassing the containment building. This event is called a steam generator tube rupture (SGTR). The quantity of radioactivity released outside of the containment building for this event would be much less than that from a containment failure for three reasons. First, the second loop is also sealed, having no direct contact with the outside environment. Second, only radioactive gases, vapors, and small particles would be able to move through the first loop into the second loop. Third, because of physical processes such as condensation and plate-out, transport through the second loop tends to remove radioactive material before it can be released to the environment.

Loss of off-site power is usually caused by factors along the electric power grid that are beyond the ability of the nuclear power plant operators to fix. The local utility companies that service the electric power grid are best equipped for this job. Restoration of on-site power from the emergency diesel generators is the next obvious step. Nuclear power plant operators have procedures and equipment for this in place and would immediately dispatch a repair crew to get the diesel generators running. Furthermore, diesel generators are tested frequently to reduce the probability of failing to start and run. However low the probability of an SBO, this scenario is a significant fraction of the risk of radiological release to the public from nuclear power plant accidents (NRC 1990).

The researchers decided to identify alternative strategies that do not rely on on-site and off-site electric power. After reviewing the literature about severe accidents, they found that many accident management strategies had already been identified for a variety of accident scenarios (see, for example Hanson et al. 1990 and Kastenberg 1991). They identified three alternatives for further study using an analysis based on decision theory: flood the reactor cavity with water, flood the reactor cavity and depressurize the reactor vessel, and do neither.

A nuclear power plant is generally equipped with an extensive fire-suppression system. The water for firefighting is contained in large tanks, containing on the order of a million gallons. Water can be pumped from these tanks through the fire suppression piping by means of diesel-driven pumps that do not require off-site and on-site electric power. The reactor vessel is located within the containment building inside a reactor cavity, which is essentially a concrete bunker that surrounds the reactor vessel. To flood the reactor cavity in the event of an SBO (the first alternative strategy), operators would pre-position a pipe from the fire-suppression tanks into the reactor cavity and fill it with water. This would retard or prevent breach of the reactor vessel and retard or prevent two modes of containment-building failure—direct heating and concrete base-mat melting— if the vessel were to breach. One disadvantage of this strategy, however, is that if the vessel breaches when it is at high pressure, an ex-vessel steam explosion would be more likely to occur.

The second alternative strategy remedies some of the disadvantages of the first. Nuclear reactors are equipped with safety valves and relief valves for the purpose of preventing overpressurization of the reactor vessel. The safety valves operate without any electric power and without operator action. The relief valves require battery power to open. The operators can manually reduce the pressure within the reactor vessel by opening one or more relief valves. In the second strategy, therefore, the relief valves are opened in addition to filling the reactor cavity. This strategy has three additional advantages. Lower pressure in the reactor vessel retards or prevents its breach; if the vessel breaches, lower pressure reduces the probability and severity of a large ex-vessel steam explosion that might endanger the containment building; and lower pressure retards or prevents SGTR.

The third strategy is to do neither of the other two, but all three strategies include all efforts to restore electric power.

10.5 The Decision Models and the Strategy Rankings

Because the consequences of a nuclear accident are also financial, the university researchers found that considerable work had been done to relate nuclear accident radiological consequences to equivalent costs. The accident at Three Mile Island Unit 2 cost the owners their investment in the plant. Additional substantial costs to the owners and their insurers were associated with cleanup, decommissioning, and lawsuits (NRC 2006c). The accident at the Chernobyl Nuclear Power Plant (about 110 km north of Kiev, Ukraine, April 26, 1986) spread radioactive contamination over a wide area, rendering losses in economic activity and costs for treating biological consequences that have been estimated in the hundreds of billions of dollars (BBC 1998).

The researchers developed four decision models using cost in dollars as the unit of consequence:

- *Perspective 1:* Model the three strategies with respect to loss of investment, in addition to costs of repair, downtime, decommissioning, cleanup, and litigation for an SBO scenario with no release of radioactivity.
- *Perspective 2:* Model the three strategies with respect to regional financial losses associated with an SBO core melt accident with subsequent release of radioactivity to the public.
- *Perspective 3:* Model the three strategies with respect to the cost of future lives saved of an SBO core melt accident with subsequent release of radioactivity to the public.
- *Perspective 4:* Model the three strategies with respect to all of these costs.

I describe how the researchers modeled each perspective in the sections that follow.

10.5.1 Perspective 1: Best Strategy with Respect to Costs of Core Damage

Figure 10-2 is the decision tree for Perspective 1. The decision node (a square) is followed by the three strategies: (1) flood reactor vessel cavity, (2) flood cavity and depressurize reactor vessel, and (3) do neither. Each strategy has a development cost of approximately $1 million. The remainder of the tree consists of chance nodes. Each chance node is associated with the probability of occurrence over 40 years of a reactor's lifetime (as a percentage above the line to the right of the circle chance node) and the cost of occurrence of each event (in dollars below the line to the right of the circle chance node). I describe the key chance events to the decision in the subsections that follow.

Core damage? This key chance event includes a partial (which happened during the Three Mile Island accident) to a full meltdown of the core. The probability of this event includes the probability of the sequence of events described in Section 10.4 up through failure to restore electric power in time to prevent core damage. This probability was from NUREG-6890 (Eide et al. 2005). Costs of future core damage accidents can be estimated only with large uncertainties. In this case, the Three Mile Island accident is a good source of information to begin the estimate. After deriving the cost of core damage from the financial costs to the owners and insurers of Three Mile Island Unit 2 (ANS 2005, 2006, NRC 2006c), the researchers estimated this cost, adjusted to 2006 dollars, at $3.6 billion. They considered this to be a median value and considered the upper bound to be a factor of two higher (i.e., the researchers had 95% confidence that the cost would be less than or equal to $7.2 billion). The estimate included costs for original construction, postaccident cleanup, decommissioning, and litigation.

Even if core damage is averted, the costs of an SBO to the owners and operators of a nuclear plant can be significant. The fuel may have exceeded its design temperature without melting, the diesel-generator repair may be time consuming, or other portions of the first or second loop may have exceeded design temperatures. As a result, extensive inspection and repair may be necessary. The researchers estimated a range of costs from an optimistic single day of outage at $1 million of lost revenue per day to as much as $500 million for an extended outage and repair.

Electric power recovered before vessel breach? This includes efforts to restore both on-site and off-site power. The probability was from NUREG/CR-6056 (Kastenberg 1993).

Vessel breach? The vessel can be breached whether electric power is recovered or not, because the power may be restored too late to prevent the breach, the operators may not have time to start up the emergency systems after electric power is restored, or the fuel configuration may not be able to be cooled. Here, the researchers derived the probabilities from a more detailed analysis of Figure 3-7 of NUREG/CR-6056 (Kastenberg 1993). If the cavity is not filled with water (i.e., for the "do neither" strategy) and electric power is not restored, vessel breach is considered a certainty. The researchers estimated the additional cost associated with cleanup and decommissioning if the melted fuel were to enter the containment building to be on the order of $1 billion.

The decision rule is to minimize the total costs of the equivalent lotteries. The costs are similar because the probability of core and vessel damage is extremely small, compensating for the large cost of such damage when the probability-weighted expected value is calculated. The expected value of each strategy is dominated by the probability-weighted cost of the occurrence of the SBO event itself, without core damage. Although all equivalent lottery costs are similar, the decision rule indicates that the "do neither" strategy would be the best choice

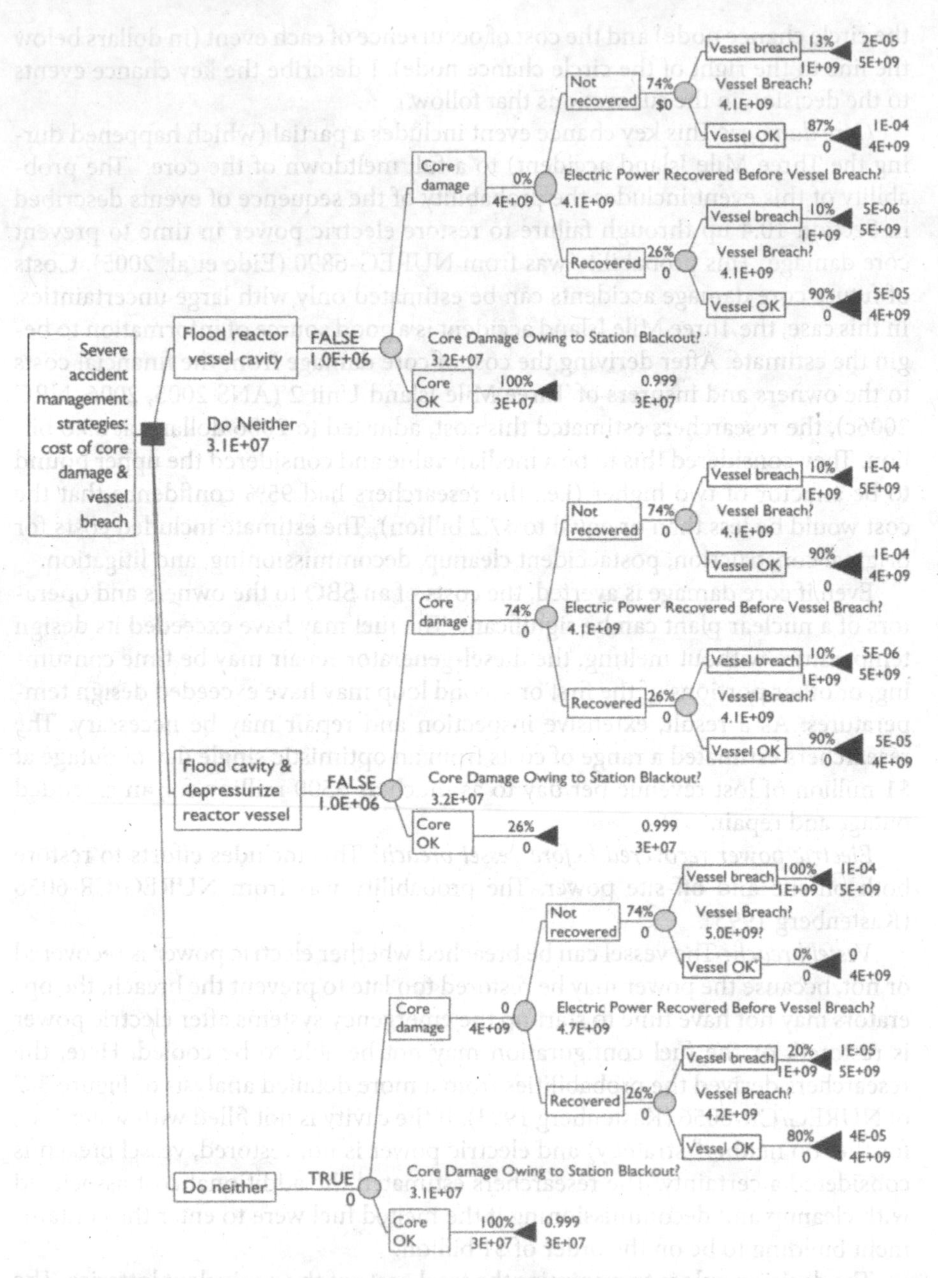

Figure 10-2. ***Decision Model for Station Blackout (SBO) Core Melt: Cost of Core Damage and Vessel Breach***

because it saves the cost of developing the strategy. A factor of 10 reduction in the core damage frequency does not change the choice of the best strategy, but a factor of 8 increase in the core damage frequency does. If the core damage frequency is at least 8 times larger than the NRC estimates, the best choice is one that minimizes the probability of vessel breach. This would be the strategy that includes both reactor cavity flooding and reactor vessel depressurization. This change in the best strategy occurs because with a sufficiently large core damage frequency, the cost of core damage and vessel breach (as opposed to the cost of the SBO event itself) has enough of an effect on the lottery to make the probability of vessel breach important to the decision.

This analysis indicated that, with the NUREG-6890-estimated SBO core damage frequency, owners and operators have little financial incentive to develop severe accident management strategies when the perspective is limited to loss of capital, cost of repair, decommissioning, downtime, cleanup, and litigation associated with core damage events that do not release radioactivity. This result, however, is sensitive to large increases in the estimated core damage frequency.

10.5.2 Perspective 2: Best Strategy with Respect to Regional Economic Costs

Figure 10-3 is an excerpt from the decision tree for Perspective 2. In this analysis, the researchers extended the decision tree in Figure 10-2 to include additional chance nodes that include containment-building failure. The probabilities and costs of the decision and chance nodes through vessel breach are the same as in Figure 10-2. The additional chance nodes are described in the following subsections.

Vessel breach? The pressure in the reactor vessel at the time of its breach influences the phenomena within the containment building and the probability of containment failure. High reactor vessel pressure at breach (generally taken to be greater than 200 psi) means that the melted fuel may eject rapidly. This may increase the likelihood of an ex-vessel steam explosion, but decrease the probability of the base-mat melt-through, because the fuel tends to disperse throughout the reactor cavity and the containment building. A high pressure in the reactor vessel, therefore, would increase the probability of vessel breach but would also reduce the subsequent conditional probability of containment failure. The strategy of flooding and reducing reactor vessel pressure would reduce the probability of vessel breach but would also increase the subsequent probability of containment failure. This same behavior could also occur if cold water began to flow into the reactor vessel before vessel failure because electric power had been restored. If there is no water in the cavity and electric power is not restored (e.g., for the do neither strategy), the vessel would breach at high pressure. The researchers derived this probability from Figure 3-7 of NUREG/CR-6056 (Kastenberg 1993).

Containment failure? The probability of containment failure depends on the path in the decision tree. A vessel breach at high pressure with water in the cavity has a lower containment failure probability than a vessel breach at low

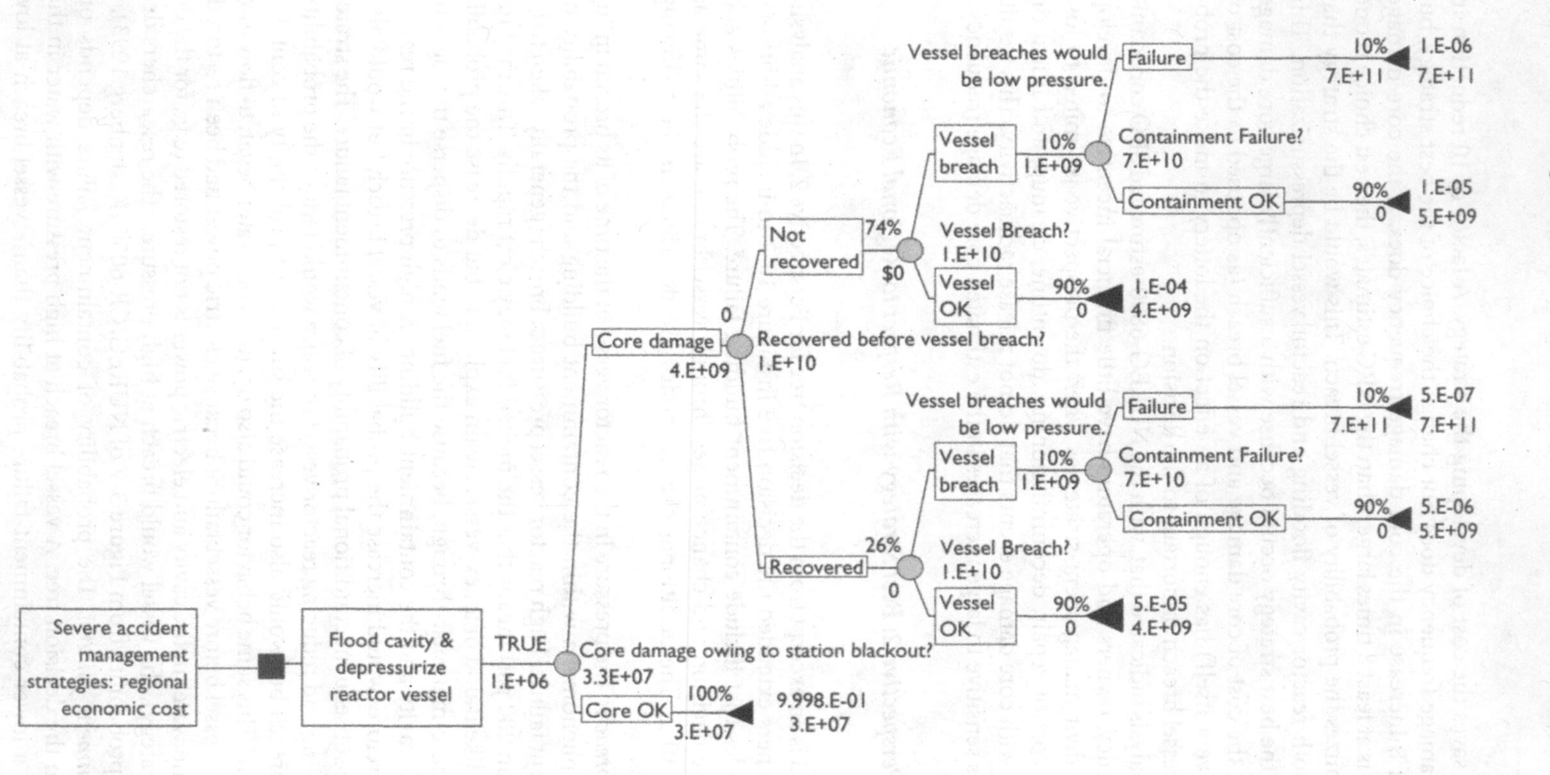

Figure 10-3. ***Excerpt of Decision Model for Station Blackout (SBO) Core Melt: Regional Costs***

pressure. Restoring electric power and the accompanying reduction in vessel pressure would reduce the probability of vessel failure. Should it fail anyway, though, the conditional probability of containment failure is higher than that of a vessel breach at high pressure. Opening the relief valves to depressurize the reactor vessel has the same effect. The do neither strategy would cause a vessel failure at high pressure. The researchers took these probabilities from Figure 3-7 of NUREG/CR-6056 (Kastenberg 1993).

This perspective investigates the local, state, and regional monetary costs of containment failure (the third perspective investigates the averted costs of lives saved). The monetary cost of containment failure and the accompanying release of radioactivity to the public is high. The best example to date from which to derive an estimate of this cost is the Chernobyl accident. The International Atomic Energy Agency (IAEA) report entitled *Chernobyl's Legacy: Health, Environmental and Socio-economic Impacts and Recommendations to the Governments of Belarus, the Russian Federation and Ukraine—The Chernobyl Forum: 2003–2005*, is a recent source of information on the economic costs (IAEA 2005).

The Chernobyl nuclear accident and government policies adopted to cope with its consequences imposed huge costs on Belarus, the Russian Federation, and Ukraine. Although these three countries bore the brunt of the impact, because radiation spread outside the borders of the Soviet Union, other countries (in Scandinavia, for instance) sustained economic losses as well.

The costs of the Chernobyl nuclear accident can be estimated only with large uncertainties because of the instability of the exchange rates and nonmarket conditions at the time. A variety of government estimates from the 1990s put the cost of the accident at hundreds of billions of dollars over two decades. Both direct and indirect costs, such as lost economic opportunity, are included in this estimate. Examples of costs are

- resettlement of people and construction of new housing and infrastructure to accommodate them;
- social protection and health care provided to the affected population;
- research on environment, health, and production of clean food;
- radiation monitoring of the environment, radioecological improvement of settlements, and disposal of radioactive waste;
- indirect losses relating to the opportunity cost of removing agricultural land and forests from use and closing agricultural and industrial facilities;
- opportunity costs, including the additional costs of energy resulting from the loss of power from the Chernobyl nuclear plant and the cancellation of the nuclear power program in Belarus; and
- lower wages and higher unemployment in the affected areas. In part, this results from the accident and its aftermath, which forced the closure of

many businesses, imposed limitations on agricultural production, added costs to product manufacture (particularly the need for constant dosimetric monitoring), and hurt marketing efforts.

It is also difficult to convert these costs into costs of a containment-building failure in the United States. On the one hand, per capita productivity and overall economic activity in the regions around nuclear power plants are higher than in the countries affected by Chernobyl. On the other hand, the Chernobyl reactor had no containment building at all. Even a failed containment building could mitigate a great deal of released radiation. Furthermore, the Chernobyl accident sparked a large fire that helped disperse radiation over a wide area. The Chernobyl type of accident cannot occur in a pressurized water reactor.

Using their judgment, the researchers estimated an upper bound cost of $2.5 trillion (i.e., 95% confidence level that the cost would be less than or equal to this value) with a median value pegged at a loss of $250 billion, which is consistent with the information about Chernobyl.

Steam generator tube rupture? The steam generator tubes could rupture in sequences that include high pressure in the reactor vessel with no restoration of electric power. For instances of containment-building failure, SGTR provides a very small amount of incremental dose. For instances of vessel failure without containment failure, an SGTR provides a very small incremental cost. This possibility is relevant, then, only for incidents of high pressure in the reactor vessel without reactor vessel failure.

The extraordinary costs associated with containment-building failure strongly favor strategies that reduce its probability. Although the equivalent lottery costs of flood reactor vessel cavity and flood cavity and depressurize reactor vessel are very close, the edge goes to the latter strategy because it is slightly better at reducing the probability of event sequences leading to containment failure. The equivalent lottery cost of the strategies that include reactor cavity flooding are approximately 60% of the cost of the do neither strategy. Stakeholders with an interest in preserving the local and regional economy have a strong incentive to select an effective accident management strategy for an SBO.

10.5.3 Perspective 3: Best Strategy with Respect to Cost of Lives Saved

The form of the decision tree of Figure 10-3 applies to this perspective as well as to Perspective 2. Here, the probabilities of the events are the same as shown in the figure but instead of economic loss, the consequences are modeled as cost of lives saved. To threaten public health, a nuclear reactor accident must release substantial amounts of radioactivity, which can occur if the containment building fails or an SGTR bypasses containment. The researchers' model of this perspective included costs for these two events. The cost of future lives saved because of technology investments, whether associated with regulatory action or not, has two components—the number of lives potentially lost because of effects of radioactivity and the cost per future life saved. Both have considerable effects of radioactivity and the cost per future life saved. Both have considerable uncertainty. The number of lives potentially lost has been estimated for the SBO scenario in NUREG-1150 (NRC 1990).

The university researchers were able to synthesize information from NUREG-1150 about radiological releases for an SBO-induced containment-building failure and SGTR with their own calculations about the mitigating effects of flooding the reactor vessel cavity. They estimated the number of potential cancer deaths associated with an SBO-induced containment-building failure to have an upper 90% confidence value of 310 with a median of 12. Estimated cancer deaths from an SGTR were estimated to be about 1/50th of these values because of the mitigating effects of transport through the second loop.

With respect to the cost of radiological release and future lives saved, the NRC has evaluated the cost per averted unit of absorbed dose for purposes of evaluating the cost benefit of its regulatory actions (see, for example, NRC 1995, Callan 1997, NRC 2004). Work in other fields has produced similar equivalent monetary units for future lives saved (see, for example, U.S. EPA 1992; Hahn 1996; Wilson 2000). Estimates range from a few hundred thousand to several millions of dollars. Lutter and colleagues (2006) developed a model of the conditions under which risk regulations that are too expensive have net adverse health effects. Two principal components of the model are the implicit value of life and the change in income associated with the change in individual behaviors (i.e., income elasticity). Using new empirical estimates for the income elasticity of many of the most consequential risk-related behaviors, Lutter's results imply that regulations that cost more than $15 million per expected life saved will have counterproductive effects on overall individual mortality.

Callan (1997) mentions a cost of averted dose accruing from average expenditures in the nuclear industry in response to the SBO regulation of $5,000 for each person-rem averted. The LD50 (i.e., the average population dose over which 50% of the exposed individuals would die) is approximately 500 rem absorbed dose.

The university researchers thought that Callan's estimate provided a basis for a reasonable median cost of life saved for an SBO accident of 500 rem × $5,000/person-rem = $2.5 million per averted death. Lutter's value of $15 million provides a good theoretical upper 90% confidence bound.

Because this perspective assigns no weight to economic costs, the strategies that most effectively prevent or reduce the consequences of containment-building failure are strongly favored. The results indicate that both severe accident management strategies that include reactor cavity flooding have similar equivalent expected costs, with the edge going to the strategy that included both reactor cavity flooding and reactor vessel pressure reduction. The equivalent costs of each of these are approximately 1/200th of the do neither strategy. Those stakeholders who have an interest in averting radiological deaths (i.e., everyone) have a strong incentive to select an effective accident management strategy for an SBO.

10.5.4 Perspective 4: Best Strategy with Respect to All Costs

Perspectives 2 and 3 favored the two strategies that included reactor cavity flooding, but Perspective 1 favored do neither. Perspective 4 combines the other three. The form of the decision model shown in Figure 10-3 also applies to this perspective. The probabilities are the same and the costs of each event are the sum of the costs of the other three studies. Because of the extraordinary costs associated with regional economic loss (Perspective 2), summing all costs produces nearly identical results to that perspective.

10.6 Propagation of Uncertainties

Estimates of probabilities associated with future SBO accidents and their costs are highly uncertain. In the university researchers' study, the uncertainties were described and quantified as probability distributions for the major events in the model. Figures 10-2 and 10-3, however, show single numbers for each event. The numbers shown are the mean values of the underlying probability distributions. To propagate the input uncertainties through each decision tree, the researchers conducted a Monte Carlo simulation. This procedure obtained a best strategy, based on cost minimization, for each of the 500 trials performed, and resulted in a distribution of best strategies for each perspective. These distributions are presented in Table 10-1.

10.7 Conclusions

In this study, the researchers created a series of decision models designed to help owners and operators of nuclear power plants choose the best severe accident management strategies. The study indicates the following:

Table 10-1. ***Best Strategy Distributions***

Perspective	*Percentage of trials selecting reactor cavity flooding*	*Percentage of trials selecting cavity flooding and depressurization of reactor vessel*	*Percentage of trials selecting do neither*
1. Core damage and vessel breach costs	0	0	100
2. Regional economic loss from containment failure	0	85	15
3. Cost of lives saved	0	100	0
4. All of the above	0	85	15

- Stakeholders interested in reducing loss of life and economic loss resulting from an SBO severe core damage accident would not select the do neither strategy. Essentially, this stakeholder group comprises everyone.
- Stakeholders interested only in minimizing the cost of core damage would select the do neither strategy if the NUREG-6890 estimate of SBO-induced core damage frequency is correct. A factor of eight or more increase in the SBO-induced core damage frequency would change the selection to one that minimizes the probability of vessel breach.
- Although input uncertainties are very large, using the mean values as inputs would obtain the same decisions as full propagation of uncertainties.

Beyond the numerical insights gained, the researchers demonstrated a powerful method for evaluating and choosing accident management strategies using the results of PRA and reasonably derived costs of consequences.

Note

1. Early severe accident investigators will recognize this scenario as TMLB' in the *Reactor Safety Study* (NRC 1975).

References

ANS (American Nuclear Society). 2005. Price Anderson Act Background Information, American Nuclear Society, November, http://www.ans.org/pi/ps/docs/ps54-bi.pdf (accessed June 12, 2007).

———. 2006. http://www.ans.org/pi/resources/sptopics/tmi/cleanup.html (accessed June 12, 2007).

BBC. 1998. http://news.bbc.co.uk/2/hi/world/monitoring/83969.stm (accessed June 12, 2007).

Callan, J. 1997. Memorandum from L. Joseph Callan to the Commissioners, Response to Staff Requirements Memorandum of May 28, 1997, Concerning Briefing on IPE Insight Report, SECY-97-180, August 6.

Eide, S.A.,C.D. Gentillon, T.E. Wierman, and D.M. Rasmuson. 2005. *Reevaluation of Station Blackout Risk at Nuclear Power Plants.* NUREG/CR-6890. Washington, DC: U.S. Nuclear Regulatory Commission (U.S. NRC).

EPRI (Electric Power Research Institute) and ERIN Engineering and Research, Inc. 1992. *Severe Accident Issue Closure Guidelines.* NUMARC 91-04. Washington, DC: Nuclear Management and Resources Council.

Hahn, R.W. 1996. *Risks, Costs, and Lives Saved: Getting Better Results from Regulation*. Washington, DC: American Enterprise Institute (AEI).

Hanson, D.J., D.W. Golden, R. Chambers, J.D. Miller, B.P. Hallbert, and C.A. Dobbe. 1990. *Depressurization as an Accident Management Strategy to Minimize the Consequences of Direct Containment Heating.* NUREG/CR-5447. Washington, DC: U.S. NRC.

Hanson, D.J., S.P. Johnson, H.S. Blackman, and M.A. Stewart. 1991. *A Systematic Process for Developing and Assessing Accident Management Plans.* NUREG/CR-5543. Washington, DC: U.S. NRC.

IAEA (International Atomic Energy Agency). 2005. *Chernobyl's Legacy: Health, Environmental and Socio-economic Impacts and Recommendations to the Governments of Belarus, the Russian Federation and Ukraine. The Chernobyl Forum: 2003–2005.* 2nd rev. http://www.iaea.org/Publications/Booklets/Chernobyl/chernobyl.pdf (accessed June 12, 2007).

Kastenberg, W.E., editor. 1991. *Summary of a Workshop on Severe Accident Management for PWRs.* NUREG/CR-5781. Washington, DC: U.S. NRC.

———. 1993. *A Framework for the Assessment of Severe Accident Management Strategies.* NUREG/CR-6056. Washington, DC: U.S. NRC.

Lutter, R., J.F. Morrall, III, and W.K. Viscusi. 2006. The Cost-per-Life-Saved Cutoff for Safety-enhancing Regulations. *Economic Inquiry* 37(4): 599. Available online at http://ei.oxfordjournals.org/cgi/content/abstract/37/4/599 (accessed .June 12, 2007).

NRC (Nuclear Regulatory Commission). 1975. *Reactor Safety Study.* NUREG 75/014 (WASH-1400). Washington, DC: U.S. NRC.

———. 1985. Policy Statement on Severe Reactor Accidents Regarding Future Designs and Existing Plants. *Federal Register* 50(153): 32138, August 8.

———. 1988. Safety Goals for the Operations of Nuclear Power Plants; Policy Statement. Title 10, *U.S. Code of Federal Regulations,* Part 50. November 30.

———. 1989. *Individual Plant Examination: Submittal Guidance.* NUREG-1335. Washington, DC: U.S. NRC.

———. 1990. *Severe Accident Risks: An Assessment for Five U.S. Nuclear Power Plants.* NUREG-1150. Washington, DC: U.S. NRC.

———. 1991. *Procedural and Submittal Guidance for the Individual Plant Examination of External Events (IPEEE) for Severe Accident Vulnerabilities.* NUREG-1407. Washington, DC: U.S. NRC.

———. 1995. *Reassessment of NRC's Dollar per Person-Rem Conversion Factor Policy.* NUREG-1530. Washington, DC: U.S. NRC.

———. 2002. Office of the General Counsel, U.S. Nuclear Regulatory Commission. Nuclear Regulatory Legislation, 107th Congress, 1st Session. NUREG-0980, 1(6): June.

———. 2004. *Regulatory Analysis Guidelines of the U.S. Nuclear Regulatory Commission.* NUREG/BR-0058 Rev. 4. Washington, DC: U.S NRC.

———. 2006a. Domestic Licensing of Production and Utilization Facilities, Title 10, *U.S. Code of Federal Regulations,* Part 50.

———. 2006b. NRC Regulatory Guides—Power Reactors (Division 1). Available online at http://www.nrc.gov/reading-rm/doc-collections/reg-guides/power-reactors/active (accessed June 12, 2007).

———. 2006c. Three Mile Island, Unit 2. Available online at http://www.nrc.gov/info-finder/decommissioning/power-reactor/three-mile-island-unit-2.html (accessed June 12, 2007).

U.S. EPA (Environmental Protection Agency). 1992. *Technical Support Document for the 1992 Citizen's Guide to Radon.* EPA 400-R-92-011. Washington, DC: U.S. EPA.

Wilson, R. 2000. Regulating Environmental Hazards. *Regulation* 23(1): 33.

CHAPTER 11

Choosing Safety: The Final Analysis

Making decisions is a fundamental life skill.
—John S. Hammond et al. (1999)

UNFORTUNATELY MOST ENGINEERS, scientists, and managers in engineering projects have not been educated in either decisionmaking or probabilistic risk assessment (PRA). Every day, though, we all make decisions that involve trade-offs among attributes such as health, career, money, and personal safety. For example, you're running late to an important meeting and you can make better time by dodging across a large busy street. You make that decision to dash across the street quickly and without much thought. So was it a good decision? If you get across the street without being injured, that's a good *outcome.* Making a good *decision,* though, doesn't depend on the outcome. Instead, it involves following a good, structured process. Following the processes I have described in this book brings structure, which helps to free decisions from being dominated by hidden goals, biases, and emotion. These processes result in more objective and cogent decisions.

11.1 Reasons to Quantify Safety

Some decisionmakers (DMs) argue that safety need not be considered in system design if all of their components, systems, and facilities comply with applicable codes and standards. But plenty of accidents with catastrophic consequences have occurred in facilities that are code- and standard-compliant. Here is just a small sample of the more famous ones:

- Feyzin (Kletz 1994)

- Space Shuttle *Challenger* (Presidential Commission on the Space Shuttle *Challenger* Accident 1986)
- Chernobyl (OECD 2002)
- Bhopal (Lapierre and Moro 2002)
- DC-10 cargo door (ASN 1996–2007)
- Flixborough (HSE 1975)

Codes and standards merely reflect an industry consensus of good practice at the time of their development. Furthermore, compilations of field data (e.g., see Denson references in Chapter 3) list tens of thousands of failures, nearly all of which occur with code- and standard-compliant equipment. Compliance with codes does not prevent equipment from failing and does not prevent disasters. In general, codes and standards attempt to define an industry-wide basis for product or system performance. Many standards in structural design and consumer products seek to reduce the likelihood and severity of injury-causing accidents. They do not ensure, however, that accidents and downtime, both major and minor, will not occur.

There are three good reasons for going beyond the codes toward a risk-management approach: avoidance of financial loss, protection against negligence during litigation, and avoidance of a catastrophic accident.

Shouldn't a stakeholder or other DM understand in a quantitative way the potential for financial loss and the magnitude of that loss? With this knowledge, will he or she not be better able to prepare for such an event? Such preparation may take the form of insurance (self or second party), facility or product operational and procedural modifications, and design changes to reduce the probability or magnitude of loss.

The second reason stems from the nearly inevitable spate of litigation that follows an accident that involves injuries. The legal concept of negligence requires vendors to provide a safe facility and safe products for reasonably foreseeable use and misuse. Prudent DMs and stakeholders would take action to demonstrate that the facility or product is safe against such events. One obvious action would be a prior attempt to foresee accident scenarios. We use PRA for this purpose.

The third reason deals with avoidance of the catastrophic events simply for the sake of avoiding serious injury, loss of life, extreme property damage, and loss of a company's good reputation. Even today, we hear that "it can't happen in our facility" or "such an event is incredible." The incredible event, however, is simply one that has not yet happened. Many postaccident investigations have shown that such events were not only foreseeable but inevitable. Again, we use PRA to anticipate and then mitigate catastrophic accidents. A couple of catastrophic accidents have occurred despite a PRA:

- Therac-25 (Leveson 1995)
- Space Shuttle *Columbia* (CAIB 2003)

Performing a PRA does not guarantee absolute safety, a state in which harm is impossible. As with making good decisions, a PRA does not guarantee a perfectly safe outcome. Following a structured and quantitative process for safety analysis, however, makes safety analysis less opinion-driven and more cogent. In a culture that nurtures safety, the burden of proof should be on those who assert that the system is safe, not on those who doubt the safety of the system. A quantitative consideration of safety during design and operational decisions goes a long way toward meeting that burden of proof. Instead of designing a system and hoping for the best, DMs involved in high-consequence, engineered systems and products should be building safety in from the beginning by actively choosing safety in concert with the methods in this book.

A PRA does for safety what a good decision process does for decisionmaking. It is axiomatic that quantitative analysis of safety in a decisionmaking context tends to make systems safer because such analysis causes more to be learned about the systems and how they can go wrong. PRA relies on both the physical principles involved in system operation and the probability that things might go wrong sometime in the future. As ably demonstrated by the following quotes, these are not new ideas.

> In physical science the first essential step in the direction of learning any subject is to find principles of numerical reckoning and practicable methods for measuring some quality connected with it. I often say that when you can measure what you are speaking about, and express it in numbers, you know something about it; but when you cannot measure it, when you cannot express it in numbers, your knowledge is of a meagre and unsatisfactory kind; it may be the beginning of knowledge, but you have scarcely in your thoughts advanced to the state of Science, whatever the matter may be.
>
> —Sir William Thompson, Lord Kelvin (1889, *80*)

> We have lower grades of knowledge, which we usually call degrees of belief, but they are really degrees of knowledge.... It may seem a strange thing to treat knowledge as a magnitude, in the same manner as length, weight, or surface. This is what all writers do who treat of probability, and what all their readers have done, long before they ever saw a book on the subject.... By degree of probability we really mean, or ought to mean, degree of belief.... Probability then, refers to and implies belief, more or less, and belief is but another name for imperfect knowledge, or it may be, expresses the mind in a state of imperfect knowledge.
>
> —A. De Morgan (1847, *171–173*)

Engineers and scientists quantify many aspects of product design, development, and operation because quantification allows us to measure how close the design or operation is to the desired requirements and provides an objective basis for comparing alternatives. Here are some selected examples of day-to-day quantification activities:

- Circuit design: voltage, current, inductance, and power
- Product aging: Arrhenius processes and L10 life
- Aircraft: lift, stress versus strength, thrust, and velocity
- Fluid systems: mass and energy transfer/transport properties and frictional losses
- Reliability: failure probability, Weibull parameters, and hazard functions

Using PRA, we can quantify safety so that it too can participate, on equal terms, with the other traditionally quantitative attributes during design tradeoffs and decisions.

11.2 Lessons from Choosing Safety

What kinds of decisions are amenable to the processes described in this book? Table 11-1 gives examples of general types of decisions. In the paragraphs following the table, I summarize and derive lessons from the decision examples and case studies I presented in this book.

A nuclear power plant pumping system was studied using the attributes of cost, unavailability, and safety (represented by the risk of latent cancers). After a quantitative analysis of the consequences of three alternatives, the DM made a common sense, logical decision based on analysis of these attributes for each of the alternatives, rather than on "gut feel" or intuition.

The project manager of an interplanetary unmanned space exploration project was inundated with ideas about improving the mission. She decided to establish decision attributes and quantify the consequences for each of the ideas, consistent with Figure 1-1. Schedule was considered a constraint. Any idea that had the potential to delay the launch was unacceptable. The other attributes were cost of development and mission failure probability. In this manner, she was

Table 11-1. ***Types of Decisions for Which PRA May Be Used with Decision Analysis***

What Decisions?	
Find best-risk reduction strategy	Improve operation, inspection, and maintenance
Go-no go	Improve design process
Improve chance of successful operation	Gain confidence that system will perform as desired
Choose best product	Prioritize critical items or scenarios
Choose design concept	Compare against a safety goal

able to manage risk during the spacecraft's design and development and quickly select the best ideas.

The management team of a new airplane door under development was nonplussed by the news that the original design did not meet regulatory safety goals and that the current engineering team was highly uncertain about costs, level of safety, and schedule in their estimates to bring it into compliance. It was the uncertainty that ultimately spurred management into action because they rightfully perceived uncertainty as risk. A more experienced engineering team was hired, and fault tree analysis was conducted to monitor compliance with the safety standards. If more than one element of the design emerged, such that both met the safety standards, the least-cost alternative was selected. When clear safety goals are articulated, the risk management process shown in Figure 1-2 applies.

A wind tunnel with the potential for a methane-air detonation was being modified to allow introduction of oxygen. Management decided that such a tunnel was needed to develop new hypersonic aircraft technology. The modification would increase the potential for detonation and a great deal of subsequent property damage if a detonation were to occur. A PRA, in which risk was defined as probability of explosion, provided the current risk of the tunnel. The PRA was then used to evaluate safety risk-reduction strategies that would compensate for the introduction of oxygen and perhaps even increase overall quantified safety. The PRA pointed the way to clear and feasible safety improvement strategies.

Management of a power plant discovered a vulnerability to flooding that had the potential of causing hazardous gas release. The DM was highly safety conscious and elected not to pursue the alternative of doing nothing, which others perceived as being the least expensive. The DM continued to seek out more alternatives and more information about those alternatives. Life-cycle costs (other than accident costs), safety risk, and operations and maintenance convenience were the attributes analyzed. Because information about a promising new technology was aggressively sought, the uncertainties in a new technology were reduced to the point that it became the preferred alternative. The benefit of the decision analysis was in guiding the effort.

A consulting firm was hired to perform a PRA on a proposed liquid propane gas tank farm. This was considered the best way to answer the simple question, "What is the risk associated with accidents?" The PRA could then be used to make safety improvements in the design as well as serve as input to the environmental impact statement for the site. The PRA showed that the best way to improve safety was to prevent a single tank burst by improving automatic tank pressure protection, improving fire brigade response, improving procedures and maintenance personnel training, and improving leak-detection capability.

The project manager of a wind tunnel whose control system was involved in redesign was confronted with spiraling increases in cost, schedule delays, and complications because of the approach taken by the engineering team in dealing with safety issues. The engineering team had performed failure mode and effects analysis (FMEA) to identify every failure mode of every component in the system that might cause a catastrophic accident. This, in itself, is not all bad. The design team, though, had continued to design modifications such as redundancy and instrumentation for each component in the FMEA. Frustrated with this, they requested that the project manager waive the safety requirements. Unwilling to do so, she recognized a decision opportunity and suggested using a more sophisticated safety technology— PRA—instead of the FMEA. In retrospect, each stakeholder in the design saw the problem differently. The design team's concern was to deliver a functioning design that met the safety requirements. The project manager's concern was to deliver a safe wind tunnel under budget. The corporate executive's concern was to provide the best and most profitable support to the company's customers. This DM recognized that the decision objectives and problem statement depend on the stakeholders. She expanded the decision objectives and problem statement to reflect all concerned. The PRA was a great help because it found that the fundamental safety concern for the control system centered on keeping the model being tested in the tunnel from hitting the inner wall surface of the tunnel. Once this was recognized, all the component-level safeguards could be eliminated in favor of an independent hard-wired watchdog controller that would stop motion of the model if it neared the wind tunnel inner surface. This was quick and easy to implement; met the objectives of all stakeholders; and resulted in a system that would be safer, more reliable, and less expensive than the previous design. The analysis was performed for a fraction of the cost savings.

The design teams of a new solid rocket motor field joint came up with new alternatives to compare to the current tang and clevis joint: an improved tang and clevis joint and a flange joint. The flange joint was subdivided into one that could be tested after rocket assembly and one that would not be tested. The objective of the designs was to improve safety (i.e., reduce the risk of hot gas leaking, similar to what caused the Challenger accident). This was an important decision because it involved high consequences both in development costs and the potential failure of future launches. Accordingly, the DM embarked on a decision analysis. In Chapter 4, I described a variety of decision analysis approaches to this situation:

- *A simple single-attribute decision tree in which cost was the only factor.* This included both out-of-pocket development cost and the expected vehicle replacement cost of accidents. The latter was calculated as a lottery associated with leak of the field joint. A PRA obtained the probability of leak, which was needed input for the expected loss associated with a launch failure caused by a field joint leak. In this treatment, safety was represented by cost of an accident.
- *A multiattribute decision tree that separated cost from safety.* Safety, in terms of probability of launch failure, was calculated using a PRA. Costs were calculated as out-of-pocket development and the expected value of vehicle replacement costs.

- *A multiattribute normative decision analysis, using a decision tree.* This included the DM's preferences in terms of cost and safety, expressed as utility functions.
- *A multiattribute descriptive decision analysis.* This used the analytic hierarchy process (AHP), which analyzes both cost and safety.
- *A utility function approach to the AHP.* This included the DM's preferences in terms of cost and safety, expressed as utility functions.

Results of the multiattribute analyses were expressed as decision trajectories that show the rankings of alternatives as a function of the DM's attitude toward safety in relation to cost. Because of differences in the mathematical approaches and the differences between using and not using utility functions, each method produced a similar but different set of decision trajectories. The differences are of secondary importance compared to the insights gained into the relative merits of each alternative from the PRA and decision analysis.

The blade-trade study was performed to assist in making a decision about which alternative wind tunnel compressor blades to purchase. The attributes of this multiattribute decision analysis were: probability of catastrophic wind tunnel damage and life-cycle costs. The management team was not willing to supply their preferences in the form of utility functions or in any other way. They wanted a quantitative analysis that used acceptable data sources within a structured method. The AHP was used with calculated life-cycle costs and probabilities of catastrophic damage as the basis for the pairwise comparison matrices. This quick, low-budget study combined PRA with cost analysis and decision analysis. It produced a useful recommendation that the management team readily followed because one option was clearly superior to the others. A rather simple PRA, which amounted to a Bayesian analysis of the available wind tunnel data and Monte Carlo simulation to combine probabilities, was sufficient for this work without the need for extensive scenario development. Choosing safety need not be a lengthy, difficult process. Extra analysis for the sake of more precision is often counterproductive.

The case study about the auxiliary power unit (APU) safety improvement strategies was performed for a high-level NASA safety manager (after the *Challenger* accident) as a demonstration of how to make safety-related decisions using PRA. It followed the so-called PRA Proof of Concept (POC) Study (MDAC 1987), which had demonstrated that PRA could usefully be applied to a space shuttle system. The NASA manager saw the need for a more structured safety decision approach that reduced the influence of preferences, judgments, attitudes, and opinion. Several different safety improvement strategies were developed from the POC study results. The event and fault trees in that study were a clear and concise basis for proposing measures that would increase safety. The decision analysis illustrated the use of PRA and cost analysis, both with the appropriate use of uncertainties, within the following decision methods:

- Maximize benefit-to-cost ratio, in which benefit is measured in reduction of loss of vehicle (LOV) frequency associated with each modification and cost is the total development and installation cost.
- Minimize expected impact, in which the expected impact (often called "expected loss") is the sum of the total development and installation cost and the expected replacement cost.
- Create a decision tree with maximization of the expected cost in which safety is converted to a cost metric (e.g., dollars).
- Create a decision tree with maximization of the weighted safety and cost utility of a risk-neutral DM.
- Follow the AHP, selecting the highest combined cost and safety ranking *without using* utility functions to establish preferences.
- Follow the AHP, selecting the highest combined cost and safety ranking *using* utility functions of a risk-neutral DM to establish preferences.
- Rely on intuition.

The actual ranking of these seven safety improvement strategies varied depending on the decision method and the decision rule. This was to be expected because (1) each method has a different mathematical combination of the attribute consequences and (2) each decision rule is an expression of preference by the DM. For example, the ranking that emerged from the expected impact method (refer to Figure 7-7) is the ranking that resulted from the AHP with risk-neutral utility functions (see Figure 7-14), provided the safety weighting factor (w_s) equals 0.55. This is also the ranking obtained using the decision tree method with risk-neutral utility functions (refer to Figure 7-16), provided that w_s equals 0.75. The expected impact method has no provision for DM preferences. The very selection of that method, however, is equivalent to a specific DM preference with respect to cost vis-à-vis safety. Other methods and other DM preferences on safety did not produce the same ranking. The intuitive approach actually belied a somewhat safety-averse attitude. This approach was found to produce a result that was equivalent to the more structured approaches only for low safety weighting factors.

The intuitive approach has another significant problem. It uses inherent judgments about decision rules and preferences at which the DM has not explicitly arrived. The DM did not recognize that he was making such judgments—they were simply the result of his implicit reasoning. When exposed, such reasoning may not be consistent with the DM's actual preferences. Using a formal decision analysis contributes to making explicit the attitudes, opinions, assumptions, and preferences that factor into a decision.

Having experienced arguments over past safety-related decisions, the DM had wanted an analysis that mitigated the effects of preferences, judgments, attitudes, and opinion. Although formal, structured decision analysis does this, he learned that no decision can be free from these. Again the important lesson here is that a formal process helps bring out these internal factors so that the decision can be made in an internally consistent and cogent manner.

A decision analysis brings all aspects of the decision into the light. This exposure of things that some DMs consider private may be uncomfortable. The decision analysis impinges on the domain and responsibilities retained by the DM. The DM has the responsibility and until he or she truly understands and trusts the decision analysis process, sharing this responsibility may be done with reluctance.

The case study about the decision to launch the *Cassini* mission was a single attribute decision performed for a go-no-go decision by the Executive Office of the President of the United States. Safety was of paramount importance to the DM at the Executive Office of the President. Historically, the risk of radioactivity release from U.S. launches is very low, and the DM wanted the *Cassini* mission not to be an exception to that history. In addition, the DM was concerned with public perceptions associated with accidents that might affect the local Florida population, no matter how small the calculated probability. He was also concerned about public perceptions of accidents that might cause a reentry of any part of the spacecraft, even if no radioactive material was released. The DM did not use a hard-and-fast rule to establish if the calculated risk was tolerably low. Instead, he wanted to be assured that a sound, scientifically defensible risk assessment had been performed and that mission risk was low in comparison to other radioactivity-related risks to which the population is exposed. The analysis was found to be scientifically defensible and comprehensive. It showed small estimated probabilities of fuel release and latent cancer fatalities and, thereby, served as sufficient reason to grant approval to launch. In this case, PRA was used successfully without the development of a formal decision and value model. Incorporation of the actual physical, chemical, meteorological, and biological phenomena involved in the PRA made it scientifically defensible.

The Micro-Met study was also a single-attribute problem that focused on developing a mission architecture that would have a high probability of mission success. It was performed with a design and engineering team who were developing a concept for a set of weather stations on Mars. I described the PRA method used, along with the sensitivity studies, in Chapter 9. The sensitivity studies helped settle the question of how many weather stations would have to be launched. Central to the analysis was the calculation of the variability associated with the reliability of future component designs. The case study showed how to use this variability to demonstrate that a good component screening process would help ensure a high-reliability mission. Conducting risk analyses at an early stage of mission design was shown to be an effective method for investigating options to reduce the estimated risk of mission failure.

University researchers developed a decision tree methodology that also used multiple chance nodes to account for alternative nuclear reactor severe accident sequence paths. Such paths differed with each selected severe accident management alternative. The example of a complete loss of electric power serving a nuclear power plant was used with three different accident management strategies: flood the containment cavity, flood the containment cavity and depressurize the reactor, and do nothing. Costs of strategy implementation, reactor downtime, core meltdown, and radioactive material releases were considered in the analysis. The analysis was performed from four perspectives, each associated with a somewhat different set of stakeholders. The perspectives were (1) obtain best strategy with respect to cost of core damage, (2) obtain best strategy with respect to regional economic costs, (3) obtain best strategy with respect to cost of lives saved, and (4) obtain best strategy with respect to all costs.

11.3 The Role of Engineers and Scientists in Making Safe Decisions

"Describing the problem, clarifying objectives, and coming up with good alternatives form the foundation of good decisions" (Hammond et al. 1999, *234*). A DM makes a good decision by recognizing the opportunity, creating a clear statement of the problem, and stating his or her objectives and the attributes against which alternatives will be evaluated. The engineers and scientists help create the alternatives and develop the decision model. As part of decisions in which safety is an important attribute, other attributes, such as cost, schedule, and performance, are also analyzed.

PRA is used to develop a safety risk model. The output of this risk model may be used directly in a normative (e.g., decision tree) or descriptive (e.g., AHP) decision model. The DM then states her values in accordance with the chosen value model (such as utility functions). Synthesizing all of this information by the mathematics of the selected model results in a ranking of alternatives. The engineers and scientists then present the results of the decision and risk assessment as a recommendation to the DM.

In Figure 1-1, all the modeling and analysis to the left of the vertical line is part of the engineers' and scientists' input to a decision. The job of the DM (e.g., management) is to review the case for and against each strategy. A DM should evaluate for reasonableness the risk analyses that produce the data for the decision model, the decision model itself, and whether the results appear consistent with her values and intuition. In essence, a DM decides on the usefulness of each alternative to her objectives in a manner consistent with her values, using the outcomes of each alternative as guidance.

In Figure 1-1, the example criteria are cost, performance, and safety. But the decision and risk methods in this book are not restricted to these and not restricted to three. For the sake of clarity, I often used cost and safety as the attributes of interest in this book because, in my experience, these are the most common attributes used when performing safety-related decision and risk analyses. The methods are valid and used for any reasonable number of attributes and alternatives. For example, the AHP has been successfully applied with 20 attributes and

6 alternatives (Saaty and Forman 1993). The exact quantities of each attribute (e.g., dollars, thrust, and failure frequency) are always uncertain because they represent potential future outcomes. Estimation of probability distributions to represent uncertainty, therefore, is not only fundamental to risk and decision analysis but increases the credibility of the analysis.

In the vast majority of decision opportunities, the process a DM uses is not formal or structured. In general, he simply uses his intuition to make a decision. In the face of multiple attributes whose outcomes are uncertain and the potential loss high, the DM's task can be daunting. It becomes the job of engineers and scientists to provide guidance on the analysis of attributes and to establish a formal process as described in this book to make sure that a good decision is made. Although making good decisions does not guarantee the desired outcome, it does increase the chance of making a desired outcome happen. Instead of designing a system and hoping for the best, DMs for high-consequence engineered systems and products can and should build safety in from the beginning. I call this choosing safety.

References

ASN (Aviation Safety Network). 1996–2007. Accident description available online at http://aviation-safety.net/database/record.php?id=19740303-1 (accessed June 11, 2007). With reference to International Civil Aviation Organization (ICAO) Circular 132-AN93: 116–125.

CAIB (*Columbia* Accident Investigation Board). 2003. Report by the *Columbia* Accident Investigation Board. http://caib.nasa.gov/ (accessed June 13, 2007).

De Morgan, A. 1847. *Formal Logic.* London: Taylor and Walton.

Hammond, J.S., R.L. Keeney, H. Raiffa. 1999. *Smart Choices*, Harvard Business School Press, Boston, MA, 2.

HSE (Health and Safety Executive).1975. The Flixborough Disaster: Report of the Court of Inquiry. HMSO, ISBN 0113610750. Available online at http://www.hse.gov.uk/comah/sragtech/caseflixboroug74.htm (accessed June 11, 2007).

Kletz, T. 1994. *What Went Wrong: Case Histories of Process Plant Disasters*, 2nd ed. Houston, TX: Gulf Publishing.

Lapierre, D., and J. Moro. 2002. *Five Past Midnight in Bhopal.* New York: Warner Books.

Leveson, N. 1995. *Safeware: System Safety and Computers.* New York: Addison-Wesley.

MDAC (McDonnell Douglas Astronautics Company). 1987. *Space Shuttle Probabilistic Risk Assessment Proof-of-Concept Study Volume III.* WP NO. 1.0-WP-VA88004-03. Houston, TX: MDAC.

OECD (Organization for Economic Cooperation and Development). 2002. *Chernobyl: An Assessment of Radiological and Health Impact: 2002 update of Chernobyl: 10 years on.* Nuclear Energy Agency, Organization for Economic Co-operation and Development, 2002.

Presidential Commission on the Space Shuttle *Challenger* Accident. 1986. *Report of the Presidential Commission on the Space Shuttle* Challenger *Accident.* Executive Order 12546, February 3. Available online at http://history.nasa.gov/rogersrep/genindex.htm (accessed June 12, 2007).

Saaty, T.L., and E.H. Forman. 1993. The Hierarchon. *The Analytic Hierarchy Process Series.* Vol. 5. Pittsburgh, PA: RWS Publications.

Thompson, Sir William, Lord Kelvin. 1889. Electrical Units of Measurement. *Popular Lectures and Addresses (PLA).* London: MacMillan.

Index